POCKET GUIDE TO
SAFETY
ESSENTIALS

SECOND EDITION

Daniel J. Snyder, M.Ed, CSP, CET, CHMM, OHST, CHST, STS

National Safety Council

NSC Press Editor: Deborah Meyer
Design and Composition: Jennifer Villarreal
Senior Director, Publications: Suzanne Powills

MISSION STATEMENT

The mission of the National Safety Council is to educate and influence society to adopt safety, health, and environmental policies, practices, and procedures that prevent and mitigate human suffering and economic losses arising from preventable causes.

COPYRIGHT, WAIVER OF FIRST SALE DOCTRINE

DISCLAIMER

Although the information and recommendations contained in this publication have been compiled from sources believed to be reliable, the National Safety Council and the Author make no guarantee as to, and assume no responsibility for, the correctness, sufficiency, or completeness of such information or recommendations. Other or additional safety measures may be required under particular circumstances.

Library of Congress Cataloging-in-Publication Data
Snyder, Daniel J.
  Pocket guide to safety essentials / J. Snyder, M.Ed, CSP, CET, CHMM.—2nd edition
     pages cm
  Includes bibliographical references.
  ISBN 978-0-87912-331-4
1. Industrial safety—United States—Handbooks, manuals, etc. I. Title.
  T55.S566 2014
  363.110973--dc23
                              2014033025
ISBN: 978-0-87912-331-4
NSC Press Product Number: 176480000
2C11232023

Dedication

This is for all those who, in their professional activities, sustain and advance the integrity, honor, and prestige of the safety, health, and environmental (SH&E) profession.

Contents

Preface

Designed as a quick reference for safety professionals, this pocket guide is a compilation of several resources related to the art and science of safety management. It is not a comprehensive text but instead is intended to lead the reader to more detailed information on the subject matter by providing highlights from a variety of safety topics. I hope that this guidebook will serve as a useful resource for anyone dedicated to the continuous improvement of occupational safety and health.

I am grateful for the continued support given to me by colleagues, friends, and most importantly my family. This guidebook is dedicated to all those committed to continuous improvement, professional development, and lifelong learning.

Acknowledgment

This second edition of the *Pocket Guide to Safety Essentials* is based on the text originally compiled by Eileen Reynolds. Thank you to Ms. Reynolds for providing a strong and useful first edition to build upon.

Thank you also to the NSC staff for compiling, proofing, and editing this Pocket Guide, and special thanks to Deborah Meyer for a superb job as project manager and my handler. Anyone who has served as my handler understands the challenges involved with this task. Deborah and her team demonstrate exemplary professionalism and serve the NSC membership well.

—Daniel J. Snyder

About the Author

Daniel J. Snyder, M.Ed, CSP, CET, CHMM, OHST, CHST, STS
With 20 years of global consulting experience, Daniel Snyder partners clients with experts to develop strategies for improving the performance of safety and health management systems through the use of gap analysis, management coaching, development of supervisors, and facilitation of a positive safety culture. As a professional speaker and author, he delivers customized, high-impact messages that are optimized for the target audience and its organizational initiatives. Mr. Snyder enjoys his opportunities to serve as a mentor to and coach of professionals seeking career advancement. He serves on several ASSE/ANSI technical standard development committees and has leadership roles in professional membership associations. He is a recognized SH&E professional who demonstrates expertise in the technical subjects of certification exam blueprints, psychometrics, human performance technology, and design of training systems. An accomplished facilitator of applied adult learning principles, he is dedicated to advancing professional development through certification exam preparation products that enable professionals to meet the challenge of illustrating their competency through education, experience, and examination. He also owns Performance Based Safety LLC and SPAN International Training LLC and is currently a doctoral candidate at the University Arkansas for Adult and Lifelong Learning. Mr. Snyder may be contacted via email at snyder@safetyconsultants.org with any questions or feedback regarding this pocket guide.

Management Theories 1

Management and production theories have been developed over the past two hundred years. Late in the nineteenth century, with the shifting emphasis from an agricultural to an industrial economy, theorists began to come into their own. Early management theorists were chiefly concerned with increasing production. More recently, theorists have become concerned with the well-being of the individual and the work group and with providing more satisfying and safer work.

Modern Principles of Safety Management
- An unsafe act, an unsafe condition, and unsafe incidents are symptoms indicating opportunities for improving the safety management system.
- Certain sets of circumstances can be predicted to produce severe injuries. These circumstances can be identified and controlled.
- Safety should be managed like any other company function. Management should direct the safety initiative by setting achievable goals and by planning, organizing, and controlling activities to achieve continuous improvement in the safety process.
- The key to effective line safety performance is management procedures that assign accountability.
- The function of safety is to constantly locate, define, and improve the operational and human conditions that allow incidents to occur by:
  - Proactively determining root causes.
  - Continuously evaluating effectiveness.
- The causes of unsafe behavior can be identified and classified. Some of the classifications are overload (improperly matching a person's capability with the load), traps, and the worker's decision to err. Each cause is one that can be controlled.
- In many cases, unsafe behavior is normal human

behavior; that is, it is the result of regular people reacting to their environment. Management's job is to change the part of the environment that leads to unsafe behavior.

- There are three major subsystems that must be dealt with in building an effective safety system: the physical, managerial, and behavioral (human factor) subsystems.
- The safety system should be appropriate to the culture of the organization.
- There is no one right way to achieve safety in an organization; however, for a safety system to be effective, it must have the following fundamental criteria:
  - Visible organizational commitment
  - Expectations for and measurement of performance at all levels
  - Flexible objectivity
  - Perceived positively
  - Evaluations for continuous performance improvement.

Frederick Taylor was one of the first recognized management theorists who deemed work deserving of systematic observation and study. A classical management theorist, he was geared toward production and developed time and motion studies. He also studied the effects of incentive on motivation, stating that humans are rational beings and make decisions based on economic rewards. Taylor also advocated the science of work, relying on scientific analysis of work and its tasks rather than the rule-of-thumb technique.

Lillian and Frank Gilbreth were contemporaries of Taylor and improved upon his technique of time and motion studies. They developed laws of human motion that evolved into their principles of motion theory, which were based on speed work achieved by mechanical innovations (an adjustable bricklayer's scaffold and cement mixers), systematic management (coordinating activities on and among construction sites, generating labor efficiency), and publicity that employed glossy pamphlets replete with photographs, many of them chronological images displaying

Frank Gilbreth's buildings in progressive stages of completion. After Frank Gilbreth's death, Lillian Gilbreth continued their work, becoming a pioneer in the field of industrial engineering. Their life is humorously portrayed in the book and movie *Cheaper by the Dozen*, which was authored by two of their 12 children.

Hans Selye researched the influence of stress on coping mechanisms and developed the General Adaptation Syndrome and its phases of alarm (fight or flight), resistance, and exhaustion.

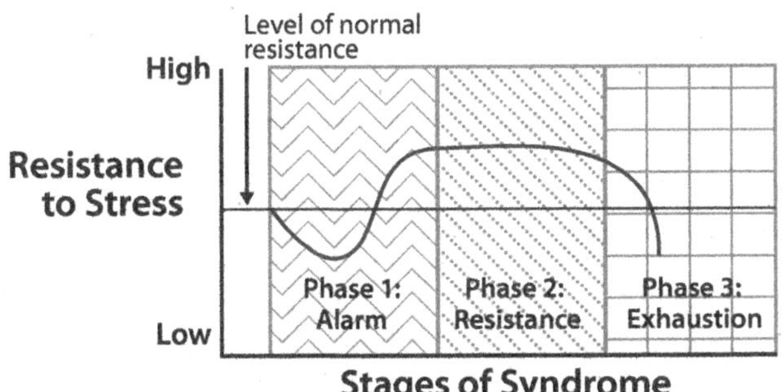

Frederick Herzberg developed the two-factor motivation hygiene theory, which describes the effects of satisfiers (motivators) and dissatisfiers (hygiene factors). Hygiene factors are elements related to the job environment, such as company policies, working conditions, salary, status, and security. Employees are not motivated by these factors, but these factors are necessary for employees to avoid dissatisfaction. Motivators are inward-generated elements, such as recognition, growth, advancement, and achievement, that are related to what people actually do on the job.

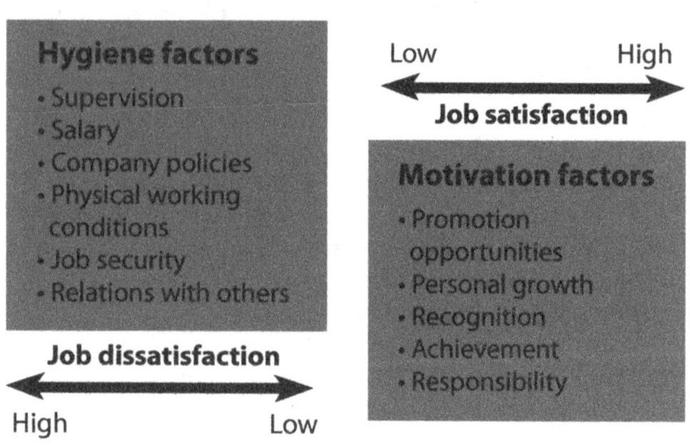

Henry Gantt emphasized the humanizing effect of management. He is probably most remembered for his technique of basing a manufacturing schedule on the time needed to complete each task—the Gantt Chart.

George Elton Mayo is remembered for his Hawthorne Studies, in which he found that people respond favorably to attention even if working conditions deteriorate. These studies were conducted at a General Electric facility in Chicago. A group of women were involved in making relay switches by hand. The experiment varied the working hours, breaks,

Gantt Chart

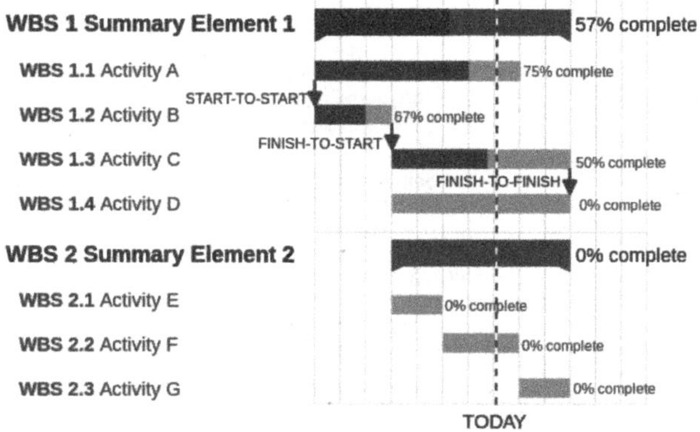

Garry L. Booker/Wikimedia Commons/Public Domain

and workweek in an attempt to find the combination that would result in increased productivity. Surprisingly, even when the working conditions were strenuous, the productivity of the group increased. The conclusions, known as the Hawthorne Effect, were that work is a group activity and that showing concern for and paying attention to the work force generally results in better work performance.

Abraham Maslow postulated the hierarchy of human needs. The needs, from lowest to highest, are physiological, safety, love, esteem, and self-actualization. When an individual's needs are fulfilled, they are replaced by needs at the next higher level. For instance, once food, clothing, and shelter are satisfied (physiological needs), then these needs are replaced by safety needs, such as protection from danger. The cycle continues until the individual reaches the level of self-actualization, where the need comes from within the individual him- or herself. Historically, organizations have used reward systems to satisfy workers at the physiological level, but current thinking is to reward workers with their needs for self-actualization.

Maslow's Hierarchy of Needs

Transcendence
helping others to
self-actualize

Self-actualization
personal growth, self-fulfilment

Aesthetic needs
beauty, balance, form, etc.

Cognitive needs
knowledge, meaning, self-awareness

Esteem needs
achievement, status, responsibility, reputation

Belongingness and Love needs
family, affection, relationships, work group, etc.

Safety needs
protection, security, order, law, limits, stability, etc.

Biological and Physiological needs
basic life needs – air, food, drink, shelter, warmth, sex, sleep, etc.

Rensis Likert, a motivational theorist, proposed the theory that both human resources and capital resources are assets that need to be capably managed. He described four management styles to illustrate his motivational theory: (1) The exploitive-authoritative system is a hierarchical system where decisions are imposed by upper management onto lower levels, resulting in little teamwork. Motivation is characterized by threats from above. Higher levels of management have great responsibilities, but lower levels have virtually no responsibilities. (2) The benevolent-authoritative system is a condescending system similar to a master-servant relationship and results in little teamwork; motivation is generally achieved because of the possibility of rewards. (3) In the consultative system, upper management has some, but not complete, trust in their subordinates. Motivation is still achieved because of the possibility of rewards, but it is supplemented by some active involvement in the process, which results in better communication and teamwork. (4) The participative-group system is the most evolved of the systems. All levels of the organization feel responsible for and are responsible for achieving goals that were decided upon in a participative process. Employees receive rewards based on their achieving these goals. Upper management has complete confidence in all levels of the structure, and the organization is rewarded by increased communication and cooperative teamwork.

Sample Likert Scale					
Indicate your opinion about the following statements using the scale: 1 = Strongly disagree. 2 = Disagree. 3 = No opinion. 4 = Agree. 5 = Strongly agree.	1	2	3	4	5
The PPE is comfortable.		X			
The management cares about my safety.			X		

Douglas McGregor wrote *The Human Side of Enterprise* in 1960, developing the idea of Theory X (work is distasteful) and Theory Y (work is play) and how organizational motives will affect the individual and, ultimately, the organization. Theory X assumes that people must be threatened to work since they do not like to work and that people do not like responsibility and must be directed. Theory X managers do not allow their employees to be fulfilled at work, and the result is not unexpected: The employees do not like their work, must be threatened, and do not take responsibility. Theory Y organizations understand that employees will be self-directed and motivated if they understand their parts in the organization's goals and are allowed to contribute to those goals; that under the right conditions, employees will seek responsibility; and that participative problem solving leads to better results than authoritative attitudes.

McGregor's XY Theory

Theory X **Theory Y**

© Businessballs.com/Alan Chapman 2002. Based on McGregor's XY-Theory. Used with permission. Retrieved Sept. 11, 2014 from Businessballs.com/mcgregorxytheorydiagram.pdf

Robert Blake and Jane Mouton developed the Blake-Mouton Managerial Grid based on a matrix of productivity and personal skills. An organization or manager with low productivity and low personal skills is considered impoverished, while an organization or manager with low productivity and high personal skills is referred to as a country club. An organization or manager with high productivity and low personal skills is considered authoritative or dictatorial, while an organization or manager with both high productivity and high personal skills is a team. The team is considered the optimal stage.

The Pareto Principle is ascribed to Vilfredo Pareto, who studied countries and their wealth. He concluded that 20% of the countries have 80% of the wealth. This has been commonly known as the 80/20 Rule. A small number of causes generate a large amount of the effect, such as 80% of a manager's time is filled by 20% of his or her problems. Looking at it from another angle, by spending

Blake-Mouton Managerial Grid

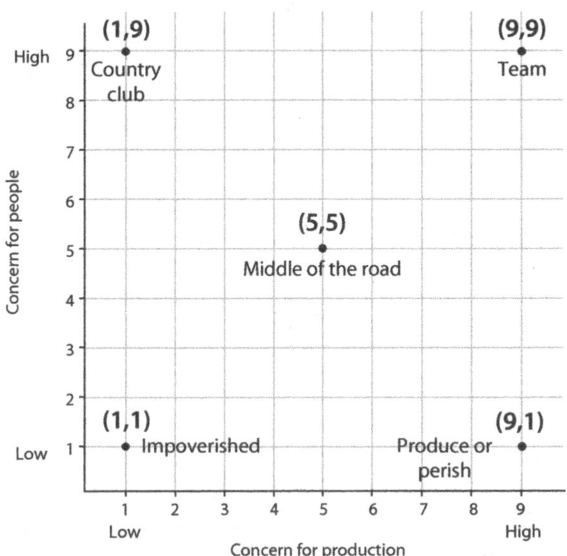

The Pareto Principle of Time versus Result

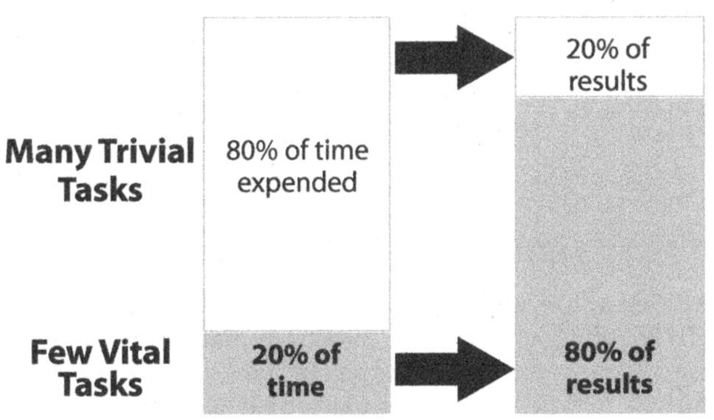

more of one's resources on a small number of the correct causes, a greater effect may be attained. The Pareto Principle was extended to the social/political realm by Joseph Juran, who said that 80% of most issues are social or political in nature, while the remaining 20% are technical. A thoroughly engineered, technical project will not solve a problem if the social/political issues are not addressed first.

David McClelland's research is concerned with the urge to achieve as a motivating factor. Financial reward is not generally a motivating factor for achievement-minded individuals; rather, the reward is the self-satisfaction of conquering the task. Financial rewards are viewed as a scorecard, a measurement device that everyone understands.

Dominant Motivator	Characteristics of This Person
Achievement	• Has a strong need to set and accomplish challenging goals. • Takes calculated risks to accomplish goals. • Likes to receive regular feedback on progress and achievements. • Often likes to work alone.
Affiliation	• Wants to belong to the group. • Wants to be liked, and will often go along with whatever the rest of the group wants to do. • Favors collaboration over competition. • Doesn't like high risk or uncertainty.
Power	• Wants to control and influence others. • Likes to win arguments. • Enjoys competition and winning. • Enjoys status and recognition.

W. Edwards Deming created a stir with his book *Out of the Crisis*, which distilled elements from many of the preceding theorists into 14 business principles. He also proposed an 85/15 Rule, which postulates that 85% of a worker's effectiveness on the job is determined by forces that he or she has no control over and that only 15% is determined by his or her skill. He also authored the well-used management systems model "Plan-Do-Check-Act" cycle.

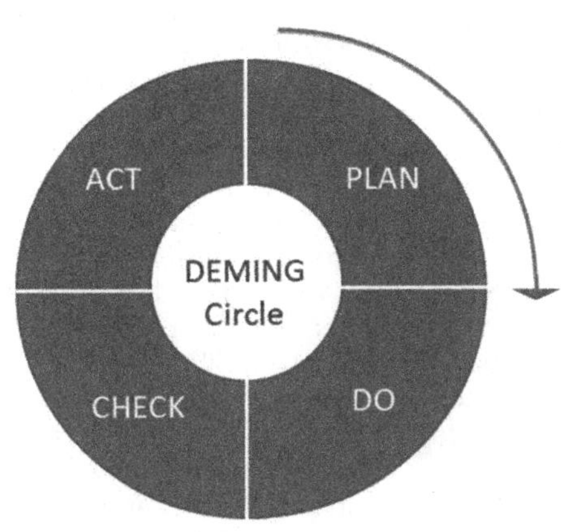

$$Quality = \frac{Results\ of\ work\ efforts}{Total\ costs}$$

Deming's 14 Principles of Management
1. Create constancy of purpose toward improvement of product and service, with the aim to become competitive, to stay in business, and to provide jobs.
2. Adopt the new philosophy. We are in a new economic age. Western management must awaken to the challenge, must learn their responsibilities, and must take on leadership for change.
3. Cease dependence on inspection to achieve quality. Eliminate the need for massive inspection by building quality into the product in the first place.
4. End the practice of awarding business on the basis of a price tag. Instead, minimize total cost. Move toward a single supplier for any one item and build a long-term relationship based on loyalty and trust.

5. Improve constantly and forever the system of production and service, to improve quality and productivity and thus constantly decrease costs.
6. Institute training on the job.
7. Institute leadership [see Point 12 and Chapter 8 of *Out of the Crisis*]. The aim of supervision should be to help people and machines and gadgets do a better job. Supervision of management is in need of overhaul, as is supervision of production workers.
8. Drive out fear so that everyone may work effectively for the company. [See Chapter 3 of *Out of the Crisis*.]
9. Break down barriers between departments. People in research, design, sales, and production must work as a team to foresee problems of production and usage that may be encountered with the product or service.
10. Eliminate slogans, exhortations, and targets for the work force that ask for zero defects and new levels of productivity. Such exhortations only create adversarial relationships, as the bulk of the causes of low quality and low productivity belong to the system and thus lie beyond the power of the work force.
    a. Eliminate work standards (quotas) on the factory floor. Substitute with leadership.
    b. Eliminate management by objective. Eliminate management by numbers and numerical goals. Instead, substitute with leadership.
11. Remove barriers that rob the hourly worker of his or her right to pride of workmanship. The responsibility of supervisors must be changed from sheer numbers to quality.
12. Remove barriers that rob people in management and in engineering of their right to pride of workmanship. This means, *inter alia*, abolishment of the annual or merit rating and of management by objectives.
13. Institute a vigorous program of education and self-improvement.
14. Put everybody in the company to work to accomplish the transformation. The transformation is everybody's job.

Dan Petersen has long maintained that safety is a "people problem." In his view, the two keys to a "world-class" safety program are to make the line organization accountable for safety performance and to have employee involvement. Yet, he asserts, the "fundamental principle" that undergirds OSHA is that "accidents are caused by things, not by people."

Dr. Petersen's 10 Basic Principles of Safety
1. An unsafe act, an unsafe condition, and an accident are all symptoms of something wrong in the management system.
2. We can predict that certain sets of circumstances will produce severe injuries. These circumstances can be identified and controlled.
3. Safety should be managed like any other company function. Management should direct the safety effort by setting achievable goals and by planning, organizing, and controlling activities to achieve them.
4. The key to effective line safety performance is management procedures that assign accountability.
5. The function of safety is to locate and define the operational errors that allow accidents to occur. This function can be carried out in two ways:
    a. by asking why accidents happen—searching for their root causes
    b. by asking whether certain known effective controls are being utilized
6. The causes of unsafe behavior can be identified and classified. Some of the classifications are overload (the improper matching of a person's capability with the load); traps; and the worker's decision to err. Each cause is one that can be controlled.
7. In most cases, unsafe behavior is normal human behavior; it is the result of regular people reacting to their environment. Management's job is to change the environment that leads to unsafe behavior.
8. There are three major subsystems that must be dealt with in building an effective safety system:
    a. the physical

    b. the managerial

    c. the behavioral

9. The safety system should be appropriate to the culture of the organization.

10. There is no one right way to achieve safety in an organization; however, for a safety system to be effective, it must meet certain criterions. The system must:
    - Force supervisory performance.
    - Involve middle management.
    - Have top management visibly showing their commitment.
    - Have employee participation.
    - Be flexible.
    - Be perceived as positive.

How Traditional Safety Programs Differ from a Systematic Safety Process

Traditional Safety Programs	Systematic Safety Process
One head office person responsible for safety, managed many times as a collateral activity.	On-the-job worker and work group responsible for incident prevention and safety goals.
An organizational safety specialist usually referred to as a safety director (sometimes safety directors have departmental support staff).	An organization dedicates more individuals to full-time rotations as safety specialists, working under the supervision of professional safety coordinator(s) or a safety administrator.
Safety as a staff function with director and staff positions.	Safety and incident prevention as a responsibility of line organization.
A set of specific Do and Don't company safety regulations set down by management. Very little focus on employee involvement.	Emphasis on upward communication from work group to supervisors to middle management to top management. Also, lateral communication to the workers' union and the workmen's compensation board as well as between work groups.

Marginal line commitment to safety and moderate involvement in accident prevention.	Umbrella policies from management protecting worker input as well as specifying structure of committees, nature of communication flow, and reward system. These policies focus on the safety process. The exact content of the rules and regulations is left to the particular work groups to formulate and enforce. (Establish safety as the group norm.)
	Management and work group involvement in and commitment to accident prevention goals and continuous improvement.

BEHAVIORAL-BASED SAFETY

Sometimes referred to as people-based safety, behavioral-based safety involves the use of observation and feedback as a metric and coaching process. Authors Tom Krause and Scott Geller have published several textbooks on the subject. Behavioral-based safety is a process, not a program, that uses applied behavior analysis methods, or a systems approach, to achieve ongoing improvement of employee safety performance.

A systems approach means that it is not possible to understand why the work force performs as it does without looking at the total safety culture of the organization. The systems approach considers issues such as:

- Training
- Equipment design
- Business and management systems
- Company values
- Peer pressure

The behavioral-based safety process requires a well-defined set of methods applied in careful sequence. The employee-driven, behavioral-based approach to safety focuses on the responsibility for ongoing peer-to-peer observation, feedback, and problem solving. This means there must be a dynamic equilibrium between management leadership and employee involvement.

Modifying or changing an existing safety culture is not necessarily an overpowering task, although it may appear to be so. By understanding that there is no singular way to best begin the behavioral-based safety process and by following established guidelines, the shift into a transformed safety culture is less stressful.

Analyze the data collected by calculating the number of safe and at-risk behaviors observed, feedback information, and other system improvements. The following is a basic formula used to determine the percent of safe behavior:

$$\text{Percent Safe Behavior} = \frac{\text{Number of Safe Observations}}{\text{Total Number of Safe and At-Risk Observations}}$$

SAFETY OBSERVATION CHECKLIST

Department ________ Date ________ Observer's Name__________

Task (Job) Observed _____________________________________

A. Personal Factors	Safe	Unsafe	E. Equipment & Tools	Safe	Unsafe
1. Rushing			1. Condition		
2. Shortcuts			2. Appropriate for Task		
3. Horseplay			3. Rigging		
4. Communication			4. Guards		
5. Knowledge of Task			5. Other		
B. Body Position			F. Vehicles		
1. Hand Placement			1. Speed		
2. Body Position			2. Passengers		

3. Footing			3. Operation		
4. Lifting/Pushing/ Pulling			4. Outriggers		
5. Other			5. Condition		
C. PPE			**G. Ladders/Scaffolds/Aerial Lifts**		
1. Hard Hat			1. Condition		
2. Safety Glasses			2. Tie-offs		
3. Hearing Protection			3. Fall Protection		
4. Safety Shoes			4. Proper Use		
5. Clothing			5. Placement		
D. Procedures			**H. Area Inspection**		
1. Confined-Space Entry			1. Lighting		
2. Welding Permits			2. Overhead Hazards		
3. Safety Rules			3. Traffic– Pedestrian		
4. Lockout/Tagout			4. Traffic–Vehicle		
5. MSDS			5. Air Quality		
6. Evacuation			6. Barricades		

I. HOUSEKEEPING OBSERVATIONS

Comments	Corrective Actions

A management system is a set of interrelated elements used to establish policy and objectives and to achieve those objectives. A management system includes organizational structure, planning activities, responsibilities, practices, procedures, processes, and resources. The conventional model for a management framework that follows a logical progression of activities aimed at improving the performance of the organization is the "Plan, Do, Check, Act" cycle.

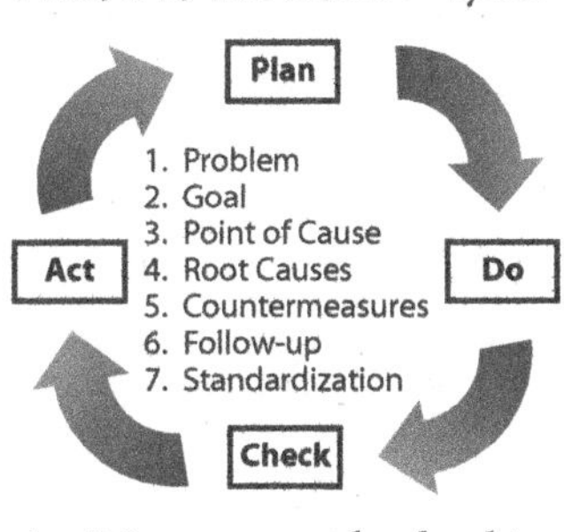

Safety and health management systems consist of six "core elements." Each core element is important and necessary to ensure the success of the overall system. All the elements are interrelated and interdependent. The core elements are:

1. Management leadership
2. Employee participation
3. Hazard identification and assessment
4. Hazard prevention and control
5. Education and training
6. System Evaluation and Improvement

An organization may choose to implement a management system for many reasons, such as enhancing business performance through the following:
- Developing a management structure that is effective and responsive to the organization's needs
- Making operational improvements
- Changing the operational culture
- Marketing opportunities and improving the operation's public image

- Improving relationships with regulators
- Enhancing the organization's ability to meet regulatory requirements and reduce costs from penalties
- Providing for greater employee involvement, awareness, and commitment to performance and improving morale
- Complying with a client's requirements

Performance Measures

<table>
<tr><td>

1. Total workers' compensation costs
2. Average cost per claim
3. Cost per man-hour
4. OSHA 200 logs
5. Industry ranking
6. Behavior observation data
7. Benchmarking other companies
8. Employee perception surveys
9. Frequency of all injuries/illnesses
10. Severity of all injuries/illnesses
11. Lost-time accidents
12. Investigations completed on time
13. Investigation identifies causes
14. Investigation identifies action plan
15. Action plan implemented
16. Safety meetings held as scheduled
17. Agenda promoted in advance
18. Safety records updated and posted
19. Inspections conducted as scheduled
20. Inspection findings brought to closure
21. Management safety communications
22. Management safety

</td><td>

   participation
23. Near miss/near hit reports
24. Discipline/violations reports
25. Self-audits for regulatory compliance
26. Contractor recordable injuries/illnesses
27. Total manufacturing process incidents
28. Total transportation incidents
29. Rate of employee suggestions/complaints
30. Resolution of suggestions/complaints
31. Vehicle accidents per mile driven
32. Safety committee initiatives
33. Management initiatives
34. Respiratory protection audit
35. Hearing conservation audit
36. Spill control audit
37. Emergency response audit
38. Toxic exposure monitoring audit
39. Ventilation audit
40. Lab safety audit
41. Health/medical services audit
42. Hazard communication audit
43. Ergonomics audit
44. Bloodborne pathogens audit

</td></tr>
</table>

45. Housekeeping audit
46. Job safety analyses
47. Lockout/tagout audit
48. Confined spaces audit
49. Machine guarding audit
50. Electrical safety audit
51. Vehicle safety audit
52. Fire protection audit
53. Employee participation rates
54. Employee housekeeping
55. Employee safety awareness
56. Employee at-risk behavior
57. Supervisor/manager participation
58. Supervisor/manager communication
59. Supervisor/manager enforcement
60. Supervisor/manager safety emphasis
61. Supervisor/manager safety awareness
62. Injury/illness cases reported on time
63. Statistical reports issued on time
64. Ratio of safety & health staff to work force
65. Safety & health spending per employee
66. Titles in safety & health library
67. Technical assistance bulletins issued
68. Policies & procedures updated on time
69. Wellness program participation rates
70. Security audits
71. Emergency drills conducted as planned
72. Percent employees trained in CPR/first aid
73. Absenteeism rates
74. Productivity per employee rates
75. Production error rates
76. Incidence of workplace violence
77. Incidence of accidental releases
78. Employee exit interviews
79. Employee focus groups
80. Community outreach initiatives
81. Off-the-job safety initiatives
82. Insurance/consultant reports
83. Reports of peer support for safety
84. Certifications of safety & health personnel
85. Percent safety goals achieved
86. Training conducted as scheduled
87. Safety training test scores
88. Statistical tracking for programs
89. Statistical process control
90. System safety analyses
91. Contractor safety activities
92. Positive reinforcement activities
93. OSHA audit—no citations
94. OSHA audit—citations, no fines
95. Willful violations
96. Serious or repeat violations
97. Other-than-serious violations
98. Total dollar amount of penalties
99. Average time to abate reported hazard
100. Average time to respond to complaint

ANSI/ASSE Z10-2012: American National Standard for Occupational Health & Safety Management Systems

This voluntary consensus standard was published by the American Society of Safety Engineers (ASSE) following American National Standards Institute (ANSI) requirements. It provides management systems with requirements and guidelines for improving occupational health and safety. Experts from labor, government, professional organizations, and industry formulated the standard after extensive examination of current national and international standards, guidelines, and practices. More information may be obtained from www.asse.org.

Occupational Health and Safety Management Systems Specification—British Standard OHSAS 18001:2007

This standard specifies requirements for an occupational health and safety management system to enable an organization to control its risks and improve its performance. The Occupational Health and Safety Assessment Series (OHSAS) Project Group, an international association of government agencies, private industries, and consulting organizations, first published the standard in 1999. Since then, there have been 16,000 certifications to the standard in more than 80 countries. The 2007 edition reflects lessons learned from users and increases its compatibility with other international safety and health management system standards and guidelines. A companion document, OHSAS 18002:2000, serves as a guide to implementing OHSAS 18001. More information may be obtained from www.bsi-global.com.

International Labor Organization Guidelines on Occupational Safety and Health Management Systems (ILO-OSH 2001)

The International Labor Organization, a United Nations agency that brings together governments, employers, and workers of its member states, has developed voluntary guidelines on safety and health management systems. These guidelines are designed as an "instrument for the development of a sustainable safety culture within the enterprise and beyond." The key elements of the guidelines are based on the concept of continuous improvement. More information may be obtained from www.ilo.org.

OSHA's Voluntary Protection Program (VPP)

The OSHA VPP recognizes and partners with businesses and worksites that demonstrate excellence in occupational safety and health. To qualify for one of the VPPs, applicants must have in place an effective safety and health management system that meets rigorous performance-based criteria. OSHA verifies qualifications through a comprehensive, on-site review process. Using one set of flexible, performance-based criteria, the VPP process emphasizes:

- Management accountability for worker safety and health
- Continual identification and elimination of hazards
- Active involvement of employees in their own protection

More information may be obtained from www.osha.gov.

OSHA's Injury and Illness Prevention Programs (I2P2)

The major elements of an effective program include:

Management Leadership

- Establish clear safety and health goals for the program and define the actions needed to achieve those goals.
- Designate one or more individuals to have overall responsibility for implementing and maintaining the program.
- Provide sufficient resources to ensure effective program implementation.

Worker Participation

- Consult with workers in developing and implementing the program and involve them in updating and evaluating the program.
- Include workers in workplace inspections and incident investigations.

- Encourage workers to report concerns, such as hazards, injuries, illnesses, and near misses.
- Protect the rights of workers who participate in the program.

Hazard Identification and Assessment
- Identify, assess, and document workplace hazards by soliciting input from workers, inspecting the workplace, and reviewing available information on hazards.
- Investigate injuries and illnesses to identify the hazards that may have caused them.
- Inform workers of the hazards in the workplace.

Hazard Prevention and Control
- Establish and implement a plan to prioritize and control hazards identified in the workplace.
- Provide interim controls to protect workers from any hazards that cannot be controlled immediately.
- Verify that all control measures are implemented and are effective.
- Discuss the hazard control plan with affected workers.

Education and Training
- Provide education and training to workers in a language and vocabulary they can understand to ensure that they know:
  - Procedures for reporting injuries, illnesses, and safety and health concerns.
  - How to recognize hazards.
  - Ways to eliminate, control, or reduce hazards.
  - Elements of the program.
  - How to participate in the program.
- Conduct refresher education and training programs periodically.

Program Evaluation and Improvement
- Conduct a periodic review of the program to determine whether it has been implemented as designed

and is making progress toward achieving its goals.

- Modify the program, as necessary, to correct deficiencies.
- Continuously look for ways to improve the program.

Steps of Data-Driven Work Group Problem Solving

"Continuous improvement" refers to the ongoing, systematic process that work groups use to examine their behavioral data and select targets for improvement. In the field of safety, the only way to be proactive is to manage by indicators that guide injury data. Ongoing behavioral data is essential because it:

- Closes the improvement loop.
- Helps identify and correct equipment and design barriers to safety.
- Establishes proactive and systematic solutions versus reactive, stopgap measures.
- Improves the safety culture through group problem solving and action planning.

Data-driven, work group problem solving makes believers of people at all levels of an organization.

1. Identifying a Problem

   The first step is to decide what to work on. There are usually several potential problems that could be improved. Effective groups don't work on everything at once. Their selection of what to work on is guided by two important criteria:
   - Action items with the most potential for reducing exposure to injury
   - Action items they can do something about

   The key to this step is using data. The group has data from observations in the forms of tables, comment reports, graphs, and injury accident reports. They have the impressions of the observers and others in the work group regarding potential problems. They may also involve themselves in specialized data collection.

2. **Identifying Root Causes**
   Having identified a performance gap, the group must determine why the gap exists. They identify the root causes, which are the basic items that need to be changed to close the performance gaps and eliminate problems. Root causes are discovered through behavior analysis and fishbone diagramming. A common mistake in problem solving is to identify a problem and then rush to develop solutions. When a group doesn't understand the causes of the problem, their solutions may miss the mark.

3. **Generating Potential Actions**
   There are usually many actions that can improve a situation. The group thinks of as many things as possible. A group working together usually can come up with more, and better, ideas than an individual working alone. Not all ideas will be equally effective, but the more ideas generated, the more likely the group can develop a good solution.

The Safety Improvement Process

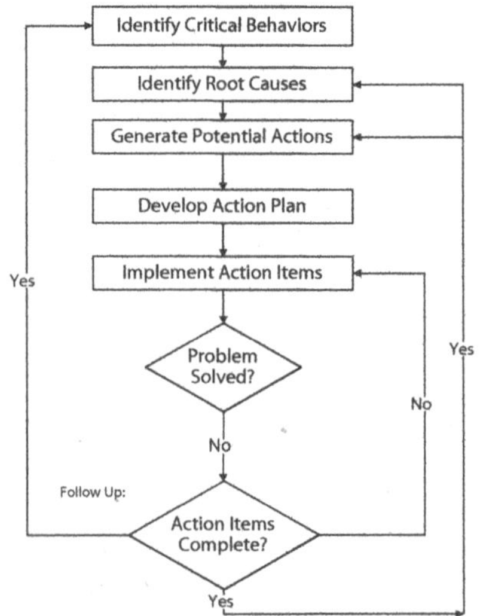

4. Evaluating Possible Actions

   Once a group has come up with potential actions, they must evaluate and prioritize them. Resources may limit the group's ability to implement all the actions simultaneously, and some actions may be inappropriate. It is thus advisable to pilot or test actions first to save money and time and to determine whether the actions work.

5. Developing an Action Plan

   From the list of possible actions, the group selects the actions they can do now. Each action and deadline is assigned to someone. This list constitutes the action plan.

6. Following Up: Measure and Evaluate

   The items in the action plan may not be the right ones to solve the problem, or they may not be implemented. So the group follows up on its action plan. They check to see whether all the action items were done. They also review ongoing behavioral observation measurements to see whether the problem they were trying to solve has actually improved. If the problem has not improved, they may need to revise the action plan.

Incidence Rates

Incidence rates can be used to show the relative level of injuries and illnesses among different industries, firms, or operations within a single firm. Because a common base and a specific period of time are involved, these rates can help determine both problem areas and progress in preventing work-related injuries and illnesses.

NOTE: When comparing illness rates by types of illness, use 20,000,000 hours instead of 200,000 hours to get a rate per 10,000 full-time employees.

An example

The following discussion illustrates how ABC Company—a fictitious construction machinery manufacturer with 200 employees—might conduct a statistical safety and health evaluation.

The ABC Company has 7 injuries and illnesses logged and 400,000 hours worked by all employees during the year. Using the formula, the incidence rate would be calculated as follows:

$$\frac{(7 \times 200{,}000)}{400{,}000} = 3.5$$

The same formula can be used to compute the incidence rate for the most serious injury and illness cases, defined here as cases that result in workers taking time off from their jobs or being transferred to another job or doing lighter (restricted) duties. ABC Company had 3 such cases.

The incidence rate for these 3 cases is computed as:

$$\frac{(3 \times 200{,}000)}{400{,}000} = 1.5$$

INCIDENCE RATE (IR) FORMULA

$$IR = \frac{[(\text{number of recordable injuries} + \text{number of recordable illnesses}) \times 200{,}000]}{\text{total hours worked}}$$

LOST WORKDAY INCIDENCE RATE (LWDI) FORMULA

$$LWDI = \frac{[(\text{number of cases away from work} + \text{number of cases restricted work}) \times 200{,}000]}{\text{total hours worked}}$$

Feasibility/Power Rating System

Purpose: Evaluate and prioritize solutions

Potential Solution	Feasibility Ratings 1 = Difficult to accomplish 2 = Moderately difficult 3 = Can be done right now	Power Ratings 1 = Short term (Quick fix) 2 = Moderately powerful 3 = Permanent solution	Overall Rating (feasibility) (power) = (overall rating)
Assigning tools to every employee	1	3	3
Training purchasing staff	2	2	4

ISO 14000 Environmental Management Systems

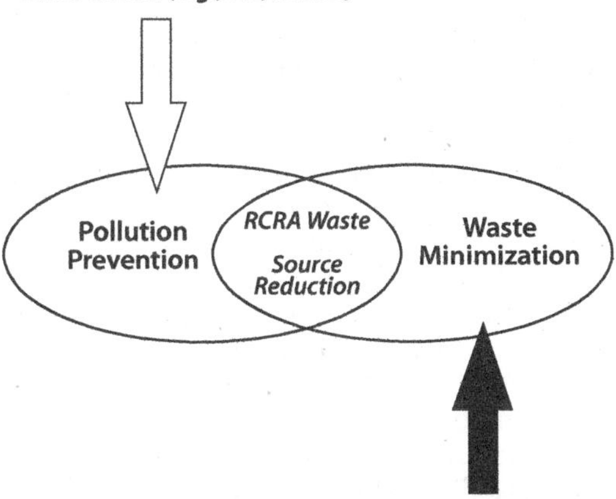

The ISO 14000 series of environmental management systems has five key components:
1. Environmental policy
2. Planning
3. Implementation and operation
4. Checking and corrective action
5. Management review

Liability, Insurance, and Compensation
- Risk Control vs. Loss Control
  - Risk Control = Preserves resources. It can prevent losses, reduce severity, and speed recovery.
  - Loss Control = Identifies hazards, assesses their risks, and establishes effective risk control measures.
- Captive = Wholly owned subsidiary of parent company
- Stock = Capitalized company; public or private
- Mutual Company = Policyholders own the company
- Lloyd's of London = Syndicate, speculative risk, high risk taker, nothing is uninsurable
- Builders' Risk = Equipment, buildings, structures, and machinery at construction site (not GL)
- Premiums:
  - Manual
  - Prospective = Based on company's past loss history and using EMR to set rates in advance
  - Retrospective = Adjust premiums at the end of the year based on actual losses during the period
- Extended Coverage (WHARVES)
  - Wind, hail, aircraft, riot, vandalism, explosion, smoke
- Pure Risk vs. Speculative Risk
  - Pure Risk = The expectation of an event that will produce only a loss should it occur
  - Speculative Risk = Result of an event that will produce a gain or loss should it occur
- Workers' Comp:
  - Whole-man Theory = Compares function of noninjured worker with ability to function after an injury. Percent disability rating is then compared to income potential.

Actions or Principles	Specific Tools	Typical Management Programs	Conceptual Management Frameworks	Overall Goal
Code of Environmental Management Principles	Material substitution	Pollution prevention	EMS	
The SIGMA Principles	Product substitution	Waste minimization	Industrial ecology	
UN Global Compact	Product reformulation	Source reduction	Design for the environment	
The Global Sullivan Principles	Process change	Cleaner production	Sigma guidelines	
Organization for Economic Cooperation and Development Guidelines for Multinational Enterprises	Equipment change	Total Quality Environmental Management	Malcolm Baldridge Quality Model	
Caux Roundtable Principles for Business	ISO 14000 family of standards	Material selection	The Balanced Scorecard	
The Principles for Global Corporate Responsibility	ISO 19011:2011	Process management	Auditing	
CERES Principles	Life cycle analysis			
Prevention of pollution	Life cycle costing Activity-based costing Material accounting Energy accounting BS8880 AA1000 Assurance Standard SA 8000 Global Reporting Initiative USD dashboard Compass index of sustainability The Natural Step Greenhouse Gas Reporting Protocol	Media program management, e.g., hazardous waste Environmental Supply Chain Management		

SUSTAINABILITY

- o Lost wages theory = Repair body parts
- o Loss of earnings capacity = Lost money, last three years' earnings average
- o Experience Modifier Rate = Adjusted actual losses/ Expected losses
- o Loss Ratio = Losses/Experience Modifier × Manual Premium
- Most Common Commercial Liability Coverage Parts Available:
  - o CGL = Comprehensive General Liability
  - o OL&T = Owners, Landlords, and Tenants
  - o M&C = Manufacturers and Contractors
  - o OCP = Owners and Contractors Protection
  - o SL = Storekeepers Liability
  - o GL = Garage Liability
- Conflict Resolutions:
  - o Arbitration = Impartial 3rd party
  - o Mediation = Impartial 3rd party; mediator has no authority to force either party to participate or agree to conditions.

LIABILITY

Liability aspects of safety are becoming more important. Terminology is complicated and detailed. Some definitions are listed.

- **Caveat emptor**—let the buyer beware
- **Caveat venditor**—let the seller beware; change from privity to strict liability; manufacturer pays for injuries
- **Contract in the mail**—in force the day it is mailed
- **Dangerous instrumentality**—anyone with a dangerous instrument is liable for any injury regardless of fault
- **Expressed warranty**—statement in writing or orally
- **Foreseeability**—held liable only if one can foresee anticipated risks
- **Hold harmless agreement**—one party assumes liability

- **Implied warranty**—dealer or manufacturer implies that product is safe
- **Liability**—obligation to rectify or recompense damage
- **Negligence**—failure to exercise reasonable care
- **Parts of contract**—agreement, parties, purpose, consideration
- **Privity**—contractual agreement directly signed between two parties; e.g., companies A, B, and C: C sues B; B sues A. No privity between A and C.
- **Product liability insurance**—protects manufacturer and retailer in case of property damage and injury
- **Reasonable care**—degree of care exercised by prudent person
- **Responsibility of handling**—care of a product in possession so that it will not later cause injury
- **Strict liability**—manufacturer liable; plaintiff doesn't need proof
- **Tort**—wrongful act or failure to exercise due care resulting in legal injury
- **Warranty of fitness**—meets buyer's use
- **Warranty of merchantability**—meets industry codes
- **WHARVES insurance**—wind, hail, aircraft, riot, vandalism, explosion, smoke coverage

COMPENSATION AND INSURANCE

Workers' compensation insurance premiums have become a significant portion of business expense. An effective method of controlling this cost is to control injuries. This is the economic basis of safety management. Compensation premiums can be calculated on a manual rate, a scheduled rate, or an experience rate.

- Experience rate—prospective, based on three years' experience (current year's rates calculated using experiences of past four, three, and two years)

The accident rates influence future premiums. Each state determines average losses by employer class.

The average rate × payroll = expected losses.

If the expected loss is exceeded, then a surcharge is assessed; if the expected loss is less, then a credit is applied.

$$\text{Loss Ratio} = \frac{\text{Losses}}{\text{E-mod} \times \text{Manual Premium}}$$

A loss ratio of 60–70% is good.

- Manual rates—rate book for each state; premiums the same from all insurance carriers; rate per $100 of payroll
- Schedule rates—reduction in rates based on hazard reduction; difficult to monitor compliance; not used currently

Organizations take several steps to prevent and control workplace hazards. On an ongoing basis, they:
- Identify and evaluate control options for workplace hazards.
- Select effective and feasible controls to eliminate, reduce, or contain these hazards.
- Implement these controls in the workplace.
- Follow up to confirm that these controls are being used and maintained properly.
- Evaluate the effectiveness of controls and improve, expand, or update them as needed.

Effective prevention and control of workplace hazards are critical to protecting employees from safety and health risks and to avoiding workplace incidents. Prevention and control allow employers to minimize or eliminate safety and health risks and liabilities as well as to meet their legal obligation to provide employees with a safe and healthy work environment. Hazard prevention and control reduce costs, improve efficiency, and boost product or service quality. They can also help improve an organization's relationships with its stakeholders and enhance its image as a responsible organization. Hazards can be prevented or controlled only after they have been identified. Therefore, most hazard prevention and control activities take place after workplace hazards have been systematically identified and assessed. Once employers understand what options are available, they can evaluate and select the most effective and feasible measures for their workplaces. This involves considering questions such as:
- What safety and health risks exist in the workplace?
- Where and how do these risks occur?
- What types of emergencies could arise, and what safety and health risks would they pose?
- What are appropriate risk reduction goals?
- What risk control technologies are available?

- How cost effective are these technologies?
- What do federal and state standards require?
- What internal standards does our organization have?
- What are current best practices within our industry?

HAZARD IDENTIFICATION

Example of Risk Assessment Matrix

	Severity and Consequences			
Likelihood of Occurrence or Exposure	**NEGLIGIBLE:** First aid or minor medical treatment	**MARGINAL:** Minor injury, lost workday	**CRITICAL:** Disability in excess of three months	**CATASTROPHIC:** Death or permanent, total disability
FREQUENT: Likely to occur repeatedly				
PROBABLE: Likely to occur several times				
OCCASIONAL: Likely to occur sometime				
REMOTE: Not likely to occur				
IMPROBABLE: Very unlikely to occur				

<u>**Risk Level:**</u>

LOW: Risk acceptable or tolerable; remedial action discretionary.
MEDIUM: Take remedial action at appropriate time.
SERIOUS: High-priority remedial action.
HIGH: Operation not permissible.

These definitions are provided for illustration purposes; an organization must define these terms as applicable to their processes.

Example of Prioritizing Scoring Method

Action Item	Frequency (Scale 1–5) 1 = High % SAFE (Less than once a week) 5 = Low % SAFE (Daily)	Severity (Scale 1–5) 1 = Least (Unlikely to result in injury) 5 = Most (Potential for fatality)	Overall Rating (Frequency) × (Severity) = Priority Rating
Use of Tools	4	4	16
Hearing Protection	3	2	6

Basic Hazard Assessment

Type of Hazard	Source Description and Guidance
Chemical agents	Safety data sheets (SDSs) provide a good basis for developing a list of toxic chemicals in the workplace. When many chemicals are present, hazards of the following types of chemicals should be determined first: chemicals that are (1) volatile; (2) handled or stored in open containers; (3) used in processes where employees are likely to be exposed through inhalation, ingestion, or skin contact; or (4) flammable and stored or used in a manner that poses a fire or explosion hazard.
Biological agents	These include bacteria, viruses, fungi, and other living organisms that can cause acute and chronic infections by entering the body either directly or through breaks in the skin. Sources can include laboratory operations, fermentation processes, or handling of raw food products. They also include exposure to blood or other body fluids or to clients or patients with infectious diseases.
Physical agents	These include excessive levels of ionizing and nonionizing electromagnetic radiation, noise, vibration, illumination, and temperature.

continued on page 38

Equipment operation	Ideally, each piece of equipment will be inspected to ensure that all safeguards necessary to protect employees are in place and effective. These include measures to ensure that employees avoid becoming caught in or struck by equipment; burned on hot surfaces; or shocked through contact with energized parts of electric circuits. Important areas of focus for equipment inspection include equipment guarding; the condition of moving parts, parts that support weight, and brakes; and hazards that might arise during maintenance activities.
Equipment maintenance	Ideally, the inspection will also include attention to safeguards to ensure that equipment maintenance can be performed safely. This would include such safeguards as de-energizing or otherwise isolating equipment; preventing chemical exposures through appropriate flushing of pumps and other process equipment; releasing stored energy; and use of lockout or tagout to prevent reactivation of equipment during servicing or maintenance.
Fire protection	This part of the inspection would include, for example, making sure that working fire extinguishers are readily available, that flammable liquids and gases are safely handled and stored, and that employees have ready access to emergency exits.
Physical environment	This involves inspecting the facility's walking and working surfaces to identify any trip and fall hazards and ensuring that they are eliminated or controlled.
Work and process flow	The flow of materials and work through an operation can be an important guide to potential hazards. For example, hazards can develop when a product produced at one stage of the process is incompatible with the equipment or practices at the next stage. This could happen in a chemical plant when one piece of equipment produces a hazardous metal catalyst packaged in 55-gallon drums that is then used 5 gallons at a time in another area of the plant. Because of this size difference, employees would need to handle the catalyst manually, causing unnecessary exposures. Producing the catalyst in 5-gallon packages would eliminate this hazard.
Work practices	Work practices can be a source of hazards. For example, inappropriate practices for lifting and handling materials can result in back and repetitive motion injuries. When potential hazards are identified, employers can consider whether employees are sufficiently trained to protect themselves. Discussing work practices with employees is particularly important as employees can often identify hazards and solutions based on their day-to-day experience with those practices.

THE HAZARD PREVENTION AND CONTROL HIERARCHY

The following are prioritized from most effective to least effective.

Controls	Examples
Elimination	Design to eliminate hazards: falls, HAZMAT, confined spaces, materials handling, tools and machinery
Substitution	Substitute for less hazardous materials and equipment, reduce energy
Engineering Controls	Incorporate safety through design such as ventilation systems, enclosures, guarding, interlocks, lift tables, conveyors
Warnings	Strategically place signs, alarms, enunciators, labels, etc.
Administrative Controls	Standard operating procedures such as job safety analyses, job rotation, inspections, training, mentoring
Personal Protective Equipment (PPE)	PPE assessments may result in the use of safety glasses, goggles, face shields, fall protection, protective footwear, gloves, respirators, chemical suits

At a minimum, organizations should implement all hazard prevention and control measures required by applicable regulatory standards. Where possible and appropriate, companies are encouraged to implement additional safety and health measures that go beyond the regulatory requirements. Once hazard prevention and control measures have been selected, they need to be implemented. The first step is to develop a written implementation plan. Implementation plans typically specify:
- What hazards need to be controlled?
- What measures will be implemented to do so?
- In what order?
- Who will implement the measures?
- By when?
- Should a written operating procedure (for example, a

standard operating procedure) be developed?
- What employee training is needed?
- When and how will implementation be confirmed?
- When and how will effectiveness be evaluated?
- When and how will routine inspections be conducted to ensure that hazard prevention and control measures remain operational?
- When and how will preventive maintenance be conducted?

Track Progress, Verify Implementation, and Evaluate Effectiveness

In an effective safety and health management system, implementation will be tracked and verified. For example:
- Have all control measures been implemented according to schedule?
- Have engineering controls been properly installed and tested?
- Have employees been trained appropriately?
- Do all employees understand the controls, including safe work practices and PPE use requirements?
- Are these controls being used correctly and consistently?

System Safety

System safety encompasses a number of disciplines. In essence, it uses logic to find out why things failed, how failure can be avoided or anticipated, how many events can be predicted in a specific period of time, and how statistics can provide objective findings. For the uninitiated, system safety is confusing and difficult, but for the adherent, it involves finding order in chaos.

When systems fail, the safety professional usually desires to find the reasons. Over the years, and especially with the development of the space program, failure analysis methods have been developed to help find the reasons. Each method has its own advantages and disadvantages. Depending upon the industry in which it is being used, one method may have significant advantages over another method.

Work Area Task Hazard Prevention and Control Worksheet

	Specific Hazard	Required PPE	Engineering Controls	Environmental Controls	Administrative Controls	Notes
Head						
Eyes/ Face						
Skin						
Hand						
Foot						
Hearing						

BOOLEAN ALGEBRA

Boolean algebra was developed by George Boole, a nineteenth-century English mathematician. It is used to express the relationships among sets or groups. In safety and health, Boolean algebra and notation are used in failure analysis methods—either in charting an event or in statistically determining the failure (or reliability) rate.

FAILURE AND RELIABILITY

Equipment components have a probability of failure, whereas events have a probability of occurrence. When components interact, the probability of failure depends on whether either component can act alone to cause the failure (either A or B) or both must fail to cause the failure (both A and B). The first instance is referred to as *union*, and, mathematically, the individual probabilities are added together for a better likelihood of occurrence. The second instance is referred to as *intersection*, and, mathematically, the individual probabilities are multiplied together for a lesser likelihood of occurrence.

$$\text{MTBF} = \text{mean time between failures} = 1/\text{failure rate}$$

Exclusive Condition
Series—reliability: Multiply probability of successes (all must work).

Parallel—failure: Multiply probability of failures (all must fail).

Probability of success $= 1 -$ probability of failure
$$P_s = 1 - P_f$$
$$P_s + P_f = 1$$
$$P_f = 1 - P_s$$

Reliability

$$\text{Component reliability} = e{-}(\lambda)(t)$$
$$\text{reliability} = e{-}(\text{failure rate})(\text{time})$$
$$\lambda = \text{failure rate}$$
$$t = \text{time; the time units must}$$

be the same in both the (failure rate) and the (time) factors

Parallel reliability—where the failure of all components is needed for system failure

$$R_{system} = 1 - (1 - R_A)(1 - R_B)(1 - R_C)...$$

$$R_A = \text{reliability of component A}$$
$$R_B = \text{reliability of component B}$$
$$R_C = \text{reliability of component C}$$

Series reliability—where the failure of any one component causes system failure

$$R_{system} = R_A \times R_B \times R_C...$$

Probabilities

- Parallel = Multiply probability of failures ("and")
- Series = Multiply probability of success ("and")
- "and" = × (multiplication)
- "or" = + (addition)

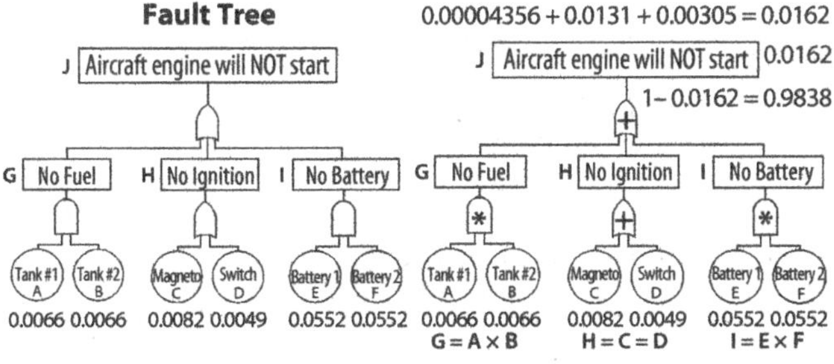

$$\text{Probability} = \frac{\text{Number of times event occurs}}{\text{Total number of possible outcomes}}$$

For this definition to be valid, each of the possible outcomes must have an equal chance of success.

AND–OR GATES

AND GATE—Multiply the probabilities together.

OR GATE—Add the probabilities together.

STATISTICS

- Central Tendency = average location of one piece of data from the set and measured using:
  - Mean = average
  - Median = middle value (middle of the road)
  - Mode = most often used value
  - Range = difference between high and low values
- Standard Deviation = how close values vary
  - One SD = 68%
  - Two SD = 95.5%
  - Three SD = 99.7%
- Coefficient of Variance = $100\sigma/\mu$ (standard deviation ÷ mean)

    Compares the percent variation of two or more groups by measures of central tendency even though measurements use different reference bases
  Samples:
  - Random—Equal and independent chance
  - Stratified—When cost is an issue
  - Systematic—By time of position (every 3rd or 5th)
  - Cluster—Small sample of a bigger group

- Coefficient of Correlation (Spearman correlation or rank correlation) = the degree of relationship between two variables
  - The coefficient of correlation expresses correlation as a number between 0 and 1 or between 0 and –1. A coefficient of 0 represents no correlation between two numbers, while a coefficient of 1 represents a strong positive correlation, with –1 representing a strong negative correlation.
  - Determines strength and direction of the relationship between two variables (age vs. salary)
  - Spurious is –6 to +6 = usually no correlation
- T-Score = compares the sample group average to the population average to determine whether the sample performs differently; also known as T-test or "Student's T"

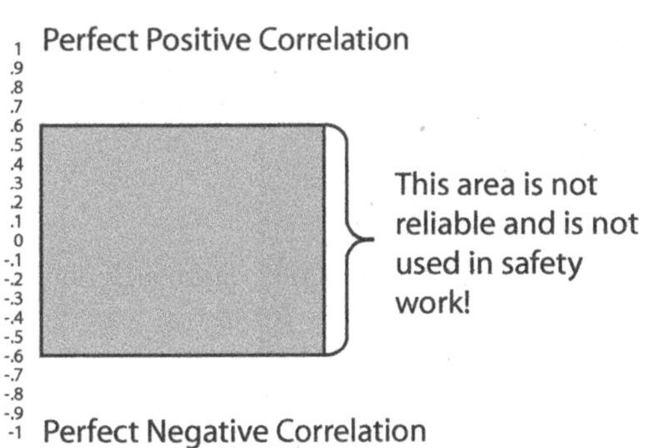

- Z-Score = determines the location of a single score in the normal distribution, the percent of the area under the normal curve, and the number of standard deviations the score is from the mean
- Chi Square = compares the theoretical and the observed values and is a test for significant difference between the frequency distributions of two groups
- Poisson Distribution = the probability that a random event (or defect) will occur a certain number of times in a given interval

FAILURE MODES

Failures happen, and the post-failure activity level can sometimes be controlled, depending on the failure mode. In the fail-passive condition, the failure causes the equipment not to function, such as when a circuit breaker trips. In the fail-active condition, the failure causes another mechanism to operate to increase the safety level. In the fail-operational condition, the failure allows certain other, critical functions, such as relief valves, to continue functioning.

Failure Analysis Methods
CCA—cause consequence analysis

CIT—critical incident technique, which is concerned with near miss interviewing

CPM—critical path, which deals with scheduling

ETA—event tree analysis, which deals with events (wanted and unwanted) and consequences (wanted and unwanted)

FHA—fault hazard analysis

FMEA—failure mode effect analysis, which is an inductive process that progresses from the specific to the general and shows the effects of failure on systems (negligible, marginal, critical, catastrophic)

FTA—fault tree analysis, which is a deductive process where the analysis begins with the event and works backward to the cause, or from the general to the specific

HAZOP—hazard and operability study, which focuses on plant operations using guide words

JSA—job safety analysis, which looks at each step of a job and the hazards associated with the job

MES—multilinear events sequencing, which is specifically geared to aircraft incidents and crashes

MORT—management oversight and risk tree, which focuses on physical equipment and management problems

PERT—program evaluation review technique, which focuses on the tasks and events in sequence with the times of occurrence and scheduling

PHA—preliminary hazard analysis, which is the initial effort to identify hazards in the design phase using tables and logic diagrams

SHA—systems hazard analysis, which deals with incompatibilities between interacting elements

SSER—system safety engineering report, which deals with the hazards of a product, its integrity, and the control of hazards

SSPP—system safety program plan, which is essentially a total safety plan

THERP—technique of human error rate prediction, which deals with human operating failures

The chemical properties of gases, liquids, and solids determine the manner in which they interact with people and the environment. This interaction may be either beneficial or detrimental, and it is necessary to determine the amount of a substance in the air or to which an individual is exposed.

A time-weighted average (TWA) is a calculation of the average exposure of an individual to a substance over a period of time. A threshold limit value (TLV®) of a chemical is the maximum amount of that chemical to which an individual may be exposed without experiencing side effects. A TLV-STEL is the short-term exposure limit, the maximum concentration of a chemical to which an individual may be exposed for 15 minutes, up to four periods per day. The TLV-C is the ceiling concentration, the exposure limit that may not be exceeded. These levels may be found in the current American Conference of Governmental Industrial Hygienists (ACGIH) publication of TLVs and biological exposure indices (BEIs).

- PEL = OSHA uses to govern worker exposure to certain materials
- STEL = short-term exposure limits (15 minutes)
- TLV® = American Conference of Governmental Industrial Hygienists developed
- TLV-C = absolute maximum exposure limit, which should never be exceeded
- Office ambient noise level = < 60 dBA
- Human hearing = 20–20,000 Hz
- Scales of measurement:
  - The A scale is very discriminatory against low frequencies.
  - The B scale is moderately discriminatory against low frequencies.
  - The C scale is almost a direct reading (it utilizes a fast response, so it is good to use as an auxiliary instrument).

Chemical and Physical Hazards Table

Hazard	Guideline		Explanation	Sources for Values
Explosion	LEL	Lower Explosive Limit	The minimum concentration of vapor in the air below which propagation of a flame will not occur in the presence of an ignition source.	NFPA
	UEL	Upper Explosive Limit	The maximum concentration of vapor in the air above which propagation of a flame will not occur in the presence of an ignition source.	NFPA
Fire	Flash Point		The lowest temperature at which the vapor of a combustible liquid can be made to ignite momentarily in the air.	NFPA
Dermal absorption of chemicals through airborne or direct contact	Designation "skin"		The designation "skin" in the ACGIH, OSHA, and NIOSH references indicates that a substance may be readily absorbed through intact skin; however, it is not a threshold for safe exposure. Direct contact with a substance designated "skin" should be avoided.	ACGIH/ OSHA/ NIOSH
Dermal irritation			Many substances irritate the skin. Consult standard references.	
Inhalation of airborne contaminants	TLV®	Threshold Limit Value	One of three categories of chemical exposure levels, defined as follows:	ACGIH
	TLV-TWA	Threshold Limit Value–Time-Weighted Average	The time-weighted average concentration for a normal 8-hour workday and a 40-hour workweek, to which nearly all workers may be repeatedly exposed without adverse effect. Should be used as an exposure guide rather than an absolute threshold.	ACGIH
	TLV-STEL	Threshold Limit Value–Short-Term Exposure Limit	A 15-minute time-weighted average exposure that should not be exceeded at any time during the workday.	ACGIH
	TLV-C	Threshold Limit Value–Ceiling	The concentration that should not be exceeded even instantaneously.	ACGIH

	PEL	Permissible Exposure Limit	Time-weighted average and ceiling concentrations similar to (and in many cases derived from) the threshold limit values published in 1968.	OSHA
	REL	Recommended Exposure Limit	Time-weighted average and ceiling concentrations based on NIOSH evaluations.	NIOSH
	IDLH	Immediately Dangerous to Life or Health	The maximum level from which a worker could escape without any escape-impairing symptoms or any irreversible health effects.	NIOSH
Carcinogens	TLV®	Threshold Limit Value	Some carcinogens have an assigned TLV.	ACGIH
	PEL	Permissible Exposure Limit	OSHA has individual standards for some specific carcinogens.	OSHA
	REL	Recommended Exposure Limit	NIOSH makes recommendations regarding exposures to carcinogens.	NIOSH
Noise	TLV®	Threshold Limit Value	Sound pressure levels and durations of exposure that represent conditions to which it is believed that nearly all workers may be repeatedly exposed without any adverse effect on their ability to hear and understand normal speech.	ACGIH
	PEL	Permissible Exposure Limit	Limits for acceptable noise exposure.	OSHA
	REL	Recommended Exposure Limit	Limits for acceptable noise exposure.	NIOSH
Ionizing Radiation		Maximum permissible body burden and maximum permissible concentration of radionuclides in air and in water.		NCRP
	PEL	Permissible Exposure Limit	Done in rems per calendar quarter.	OSHA

- Noise rules

Duration per Day (hr)	Maximum Decibels
8	90
4	95
2	100
1	105
0.5	110
0.25	115
Impact	140

Reference: After OSHA 1910.25 Table G16

  - Reduce noise sources by half; subtract 3 dBA from the initial noise source.
  - Reduce noise exposure time by half; subtract 5 dBA from allowed exposure.
  - Double the distance from the noise source; subtract 6 dBA from the noise at the source.
- Hand movement is 63 inches per second (guarding)
- Air cleaning of gases and vapors
- Explosimeter or combustible gas indicator operates on "heat of combustion"
- Adsorption = takes place on the surface of the adsorbent material, where gas and solid make contact
- Absorption = dissolves or reacts chemically with a gas or vapor that is to be removed
- Combustion = used for removal if gas or vapor can be oxidized (flame afterburner)
- Condensation = process of lowering temperature of the incoming air to change the vapor to a liquid state
- Walten-Beckett Graticule = equipment used to count asbestos fibers by phase contrast microscopy
- Air gap = physical separation between potable and not potable water to prevent contamination
- The 15-minute personal monitoring sampling strategy is good for determining exposures above the ceiling concentration

- Sampling methods:
  - Colorimetric measurement tubes = often used because they are cheap and easy to read and have a long shelf life. Produce color change with intensity.
  - Gavimetric = weighs samples that are gathered.
  - Volumetric = considers the reaction that is produced when a sample is mixed with a specific volume of an indicated agent.
  - Gas chromatography = determines how long it takes for a substance to move through a specific material designed for detection purposes.
  - Spectroscopy = involves the creation of a spectrum that produces a unique characteristic if certain contaminants are present by using a carbon arc, electron beam, or infrared radiation.
- TLVs refer to airborne concentrations (not skin contact) that nearly all workers may be repeatedly exposed to and are based on the best available information from industrial experience, human/animal studies, and freedom from irritation.

TLV CALCULATIONS

To determine if an individual has been overexposed to a substance, calculate using the exposure and TLV® in the following equation. This equation may be used to determine the TLV® of gaseous mixtures or different chemicals with the same route of entry.

The TLV® for a gas mixture may be calculated using the following equation:

$TLV_{mix} = C_1/TLV_1 + C_2/TLV_2 + C_3/TLV_3....$ If the result is greater than 1.0, then the individual has been overexposed (exposure to gas_1/allowed + exposure to gas_2/allowed...). If the result is less than 1.0, then the exposure is within the allowed limits.

The TLV® (in parts per million, ppm) for a liquid mixture may be calculated from this equation, where the percentage is expressed in decimal form:

$$TLV_{liquid\ mix} = 1/[(\%_{liquid\ 1}/TLV_{liquid\ 1}) + (\%_{liquid\ 2}/TLV_{liquid\ 2})]$$

For example, if a liquid is 40% liquid 1 with a 500 ppm TLV® and 60% liquid 2 with a 300 ppm TLV®, the TLV® for the mixture is:

$1/[(40\% \times 100/500) + (60\% \times 100/300)]$
$= 1/(0.4/500 + 0.6/300)$
$= 357$ ppm for mixture

The calculation for ppm can be based on the TLV® of the chemical or on the concentration of the chemical in air.

ppm = [(TLV mg/m^3)(24.45)]/molecular weight

Example: acetone at a concentration of 1,780 mg/m^3
$(CH_3)_2CO$ (12 + 3)(2) + (12 + 16)
$= 58$ molecular weight
$= (1,780 \times 24.45)/58$
$= 750$ ppm

Likewise, if the concentration is in ppm, then the conversion to mg/m^3 is calculated by using the following formula:

mg/m^3 = [(TLV in ppm)(molecular weight)]/24.45

TWA CALCULATIONS

TWA $= C_1T_1 + C_2T_2....$ /total time
C_1 = concentration of substance 1
T_1 = time exposed to substance 1

HEARING AND NOISE

Sound is basically a wave that travels from a source to a recipient. How it travels depends on temperature, humidity, other sources of sound, and the condition of the recipient.

Understanding sound begins with understanding the physics of sound and then evolves into understanding the sound's basic physical properties.

Amplitude = intensity, maximum height value of

sound wave, related to loudness; expressed as force/unit area, dynes/cm^2, newtons/m^2

Audiogram first detects 3,000–6,000 Hz.

A scale—1,000–2,500 Hz, mimics human ear

B scale—500–1,000 Hz

C scale—30–8,000 Hz

Average hearing range = 20–20,000 Hz

Average office sound level = 60 dBA

Continuous noise = 115 dB for 1 minute

Decibel (dB) = logarithmic unit expressing the sound pressure level

Double the distance apart = reduces dB by 6 (Rule of 6)

Double the sound pressure = adds 6 dB

Frequency (f) = cycles/second = hertz = 1/time = sound speed/wavelength, related to pitch

Greater the intensity, the higher the frequency

Hearing damage = 1,000–4,000 Hz

Impact noise = less than 1 second long and at least 1 second apart

Integrated sound level meter (SLM) = average sound level over a certain period of time

Intensity = energy/unit time, joules/(sec/m^2)

Noise reduction coefficient (NRC) = average of absorption coefficients at 250, 500, 1,000, and 2,000 Hz

Noise reduction rating = NRR

NRR of 6 = reduction of 1.0 dB

NRR (under lab conditions)—7 = effective field NRR

Period = time for wave cycle, cycles/time, 1/frequency

Sabin = sound absorption unit

Sound absorption coefficient = the amount absorbed (not reflected)

Sound level meter (SLM) = instantaneous reading of sound level

Sound power = watts

Spectrum analyzer = distribution of sound frequencies, octave bands

Speech = 500–2,000 Hz

Time = 1/frequency

Wavelength (λ) = distance for 1 cycle

Formulas

Noise reduction by absorption:

$NR = 10 \log A_2/A_1$

NR = noise reduction in dB

A_1 = total absorption units before treatment, in Sabins

A_2 = total absorption units after treatment, in Sabins

Noise reduction by distance:

$dB_1 = dB_0 + 20 \log d_0/d_1$

dB_0 = noise level at distance d_0

dB_1 = noise level at distance d_1

Rule of thumb: Double distance lowers dB level by 6.

d_0 = the reference point

d_1 = the new location

Noise reduction through a duct:

$NR \text{ duct} = (12.6 \ Pa^{1.4}/A$ dB per foot of length

NR = noise reduction in dB per foot

P = duct perimeter in inches

α = absorption coefficient

A = duct cross-sectional area in square inches

Sound intensity level:

$L_1 = 10 \log I/I_0 \ dB = 10 \log I + 120$

I = intensity of measured sound level in watts per m^2

I_0 = the reference sound intensity, generally 10^{-12}
watts/m^2

Sound pressure level:

$SPL \text{ total} = SPL_{individual} + 10 \log (n)$

n = number of sources of noise

Sound pressure level:

$L_p = 20 \log p/p_0 = 20 \log_p + 94$

L_p = sound pressure level

p = measured RMS sound pressure in N per m^2

p_0 = threshold of human hearing (0.00002 N/m^2 or
2×10^{-5} N per m^2 or 20 micropascals or 2×10^{-5} pascals)

From this equation, decibels are determined.

Sound transmission through a material:
Sound transmission loss = $\log_{10} E_t/E_i$
  E_t = transmitted sound energy
  E_i = incident sound energy

Speed of sound calculations:
  $c = \lambda \times f$
  c = speed of sound
  λ = wavelength
  f = frequency

  $c = 49(T_a + 460)^{0.5}$
  T_a = temperature of air in °F
  c = sound speed in ft/sec

Speed of sound in air = 1,130 ft/sec at 70°F
Speed of sound in water = 4,700 ft/sec
Speed of sound in metals = 16,000 ft/sec

Occupational Noise Exposure Limits

Because of the physical properties of sound and its effects on the human ear, the Occupational Safety and Health Administration (OSHA) has developed standardized guidelines for occupational exposure to sound (noise). A worker is allowed exposure to only a set limit of noise, measured in decibels. In work environments with higher exposure levels, exposure can be limited through the use of insulation or sound absorbing devices at the source or personal protective equipment used at the recipient site.

Noise exposure can be calculated by using several different equations, depending on what is known about the exposure. It should be noted that OSHA considers the noise level doubled at a difference of 5 dB, while the American Conference of Governmental Industrial Hygienists (ACGIH) considers the noise level doubled at 3 dB.

A-Weighted Sound Level (L, in decibels)	Reference Duration T (hour)
80	32
81	27.9
82	24.3
83	21.1
84	18.4
85	16
86	13.9
87	12.1
88	10.6
89	9.2
90	8
91	7.0
92	6.1
93	5.3
94	4.6
95	4
96	3.5
97	3.0
98	2.6
99	2.3
100	2
101	1.7
102	1.5
103	1.3
104	1.1
105	1
106	0.87
107	0.76
108	0.66
109	0.57

110	0.5
111	0.44
112	0.38
113	0.33
114	0.29
115	0.25
116	0.22
117	0.19
118	0.16
119	0.14
120	0.125
121	0.11
122	0.095
123	0.082
124	0.072
125	0.063
126	0.054
127	0.047
128	0.041
129	0.036
130	0.031

Amount of time a worker can be exposed to continuous noise—maximum time period for exposure at a known dB level:

$$T = 8/[2^{(L-90)/5}]$$

$$\text{time limit} = 8/[2^{(\text{sound in dB} - 90)/5}]$$

T = time limit of exposure at L dB (hours)

L = sound level in dB

Use of this formula leads to the following chart:

Duration per Day (hr)	Maximum Decibels
8	90
4	95
2	100
1	105
0.5	110
0.25	115
Impact	140

Reference: After OSHA 1910.25 Table G16

Additive noise calculations:

xdB $+ x$dB $= (x + 3)$dB doubled intensity
At $2x$ the distance: ½ noise level
At ½x the distance: $2x$ noise level
Arrange consecutively and keep rolling over.
Arrange in ascending order and compare first to second and find resulting dB$_1$, compare dB$_1$ to third and find resulting dB$_2$, etc., to final absolute resultant, which may be higher than highest individual sound level.

Difference Between the Two Levels	Additive to Higher dB Level
0	3.0
1	2.6
2	2.1
3	1.8
4	1.4
5	1.2
6	1.0
7	0.8
8	0.6
9	0.5
10	0.4
11	0.3
12	0.2

Reference: After Marshall, Brauer

Effective daily dose:

$$D = C_1/T_1 + C_2/T_2 + C_3/T_3....$$

D = dose

C_1 = length of time exposed to a certain dB level

T_1 = length of time allowed to be exposed to that level

C_2 = length of time exposed to a second dB level

T_2 = length of time allowed to be exposed to a second dB level

If total is less than 1, the worker has not been exposed to a harmful amount of noise.

Multiplying the result by 100 indicates the percent dose (%).

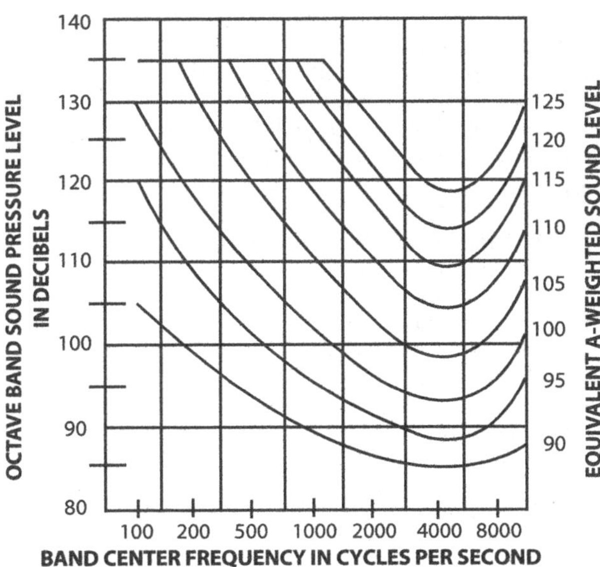

Band Center Frequency in Cycles per Second

TWA calculated from % dose:

TWA = 16.61 log (D/100) + 90

TWA = equivalent sound pressure level in dB

D = noise dose over an 8-hour day in %

ERGONOMICS

The numbers of ergonomic injuries and illnesses have been increasing in recent years. This may be due to a combination of better reporting, better diagnosis, and increased incidence. The prevention of these incidents is rapidly becoming more important to both businesses and individuals. Also of concern are heat exposure, cold exposure, and windchill.

NIOSH Lifting Equations

The recommended weight limit (RWL) calculation is used to determine whether a particular load is too heavy for an individual to lift safely, based on frequency, height of lift, asymmetry, distance from floor, coupling, and horizontal distance. If the weight of the object is known, then the calculation is solved for the lift constant (LC). The goal is for the lift constant to be less than 1. The calculation can also be worked backward, using a lift constant of 1, to determine the maximum weight of an object to be lifted, the maximum frequency of lift, or any of the other factors.

$$RWL = LC \times HM \times VM \times DM \times AM \times FM \times CM$$

RWL = recommended weight limit
LC = lift constant
HM = horizontal measure
VM = vertical measure from floor
DM = distance lifted measure
AM = asymmetry measure
FM = frequency measure
CM = coupling measure

$$RWL\ kg = (23)(25/HM)[1 - (0.003|VM - 75|)][0.82 + (4.5/DM)](1 - 0.0032AM)(FM)(CM)$$

$$RWL\ lb = (51)(10/HM)[1 - (0.0075|VM - 30|)][0.82 + (1.8/DM)](1 - 0.0032AM)(FM)(CM)$$

LI = L/RWL
LI = lift index
L = load
RWL = recommended weight limit
Looking for the LI to be less than 1

The National Safety Council and NIOSH offer good ergonomic lifting calculators, which make it quick and easy to determine the lift index.

Radiation

$$R = 11.3(T_w - T_s)$$

R = radiation in kcal/hour

T_w = mean radiant temperature of solid surroundings in °C

T_s = skin temperature in °C

If T_w is higher than T_s, then heat is added (+R).

If T_s is higher than T_w, then heat is lost (–R).

Surface Area of Body

$$SA = 71.84 \times 10^{-4}(W^{0.425})(H^{0.725})$$

SA = surface area of body

W = weight in kilograms

H = height in centimeters

Windchill

The windchill calculation and chart show the "feels like" skin temperature as based on windspeed and temperature.

Cooling Power of Wind on Exposed Flesh Expressed as Equivalent Temperature (under calm conditions)*

Estimated Wind Speed (in mph)	Actual Temperature Reading (°F)											
	50	40	30	20	10	0	-10	-20	-30	-40	-50	-60
	Equivalent Chill Temperature											
Calm	50	40	30	20	10	0	-10	-20	-30	-40	-50	-60
5	48	37	27	16	6	-5	-15	-26	-36	-47	-57	-68
10	40	28	16	4	-9	-24	-33	-46	-58	-70	-83	-95
15	36	22	9	-5	-18	-32	-45	-58	-72	-85	-99	-112
20	32	18	4	-10	-25	-39	-53	-67	-82	-96	-110	-121
25	30	16	0	-15	-29	-44	-59	-74	-88	-104	-118	-133
30	28	13	-2	-18	-33	-48	-63	-79	-94	-109	-125	-140
35	27	11	-4	-20	-35	-51	-67	-82	-98	-113	-129	-145
40	26	10	-6	-21	-37	-53	-69	-85	-100	-116	-132	-148

(Wind speeds greater than 40 mph have little additional effect.)	LITTLE DANGER In < 1 hr with dry skin. Maximum danger of false sense of security.	INCREASING DANGER Danger from freezing of exposed flesh within one minute.	GREAT DANGER Flesh may freeze within 30 seconds.
	Trench foot and immersion foot may occur at any point on this chart.		

*Developed by U.S. Army Research Institute of Environmental Medicine, Natick, MA

■ Equivalent chill temperature requiring dry clothing to maintain core body temperature above 36°C (98.8°F) per cold stress TLV®.

Windchill = $[(100V)^{0.5} - V + 10.45] \times (33 - T_a)$
Windchill in kcal/m²/hr
  V = windspeed in meters/second
 33 = skin temperature in °C
 T_a = air temperature in °C

An online windchill calculator can be found at www.nws. noaa.gov/om/windchill/.

Common Types of Cold Stress
- Hypothermia
  - Normal body temperature (98.6°F) drops to 95°F or less
  - Mild symptoms: alert but shivering
  - Moderate to severe symptoms: shivering stops; confusion; slurred speech; heart rate/breathing slow; loss of consciousness; death results
- Frostbite
  - Body tissues—e.g., hands and feet—freeze. Can occur at temperatures above freezing due to windchill; may necessitate amputation
  - Symptoms: numbness, reddened skin develops gray/white patches, feels firm/hard, and may blister
- Trench Foot (also known as Immersion Foot)
  - Nonfreezing injury to the foot caused by lengthy exposure to wet and cold environment; can occur at air temperatures as high as 60°F if feet are constantly wet
  - Symptoms: redness, swelling, numbness, and blisters
- Risk Factors
  - Dressing improperly, wet clothing/skin, and exhaustion
- For Prevention, Your Employer Should:
  - Train you on cold stress hazards and prevention
  - Provide engineering controls—e.g., radiant heaters
  - Gradually introduce workers to the cold; monitor workers; schedule breaks in warm areas

Work/Warm-up Schedule for a 4-hour Shift

Air Temperature–Sunny Sky		No Noticeable Wind		5 mph Wind		10 mph Wind		15 mph Wind		20 mph Wind	
°C (approximate)	°F (approximate)	Maximum Work Period	Number of Breaks	Maximum Work Period	Number of Breaks	Maximum Work Period	Number of Breaks	Maximum Work Period	Number of Breaks	Maximum Work Period	Number of Breaks
-26 to -28	-15 to -19	(Normal Breaks) 1		(Normal Breaks) 1		75 min	2	55 min	3	40 min	4
-29 to -31	-20 to -24	(Normal Breaks) 1		75 min	2	55 min	3	40 min	4	30 min	5
-32 to -34	-25 to -29	75 min	2	55 min	3	40 min	4	30 min	5	Non-emergency work should cease	
-35 to -37	-30 to -34	55 min	3	40 min	4	30 min	5	Non-emergency work should cease			
-38 to -39	-35 to -39	40 min	4	30 min	5	Non-emergency work should cease					
-40 to -42	-40 to -44	30 min	5	Non-emergency work should cease							
-43 & below	-45 & below	Non-emergency work should cease									

Schedule applies to any 4-hour work period with moderate to heavy work activity; with warm-up periods of ten (10) minutes in a warm location and with an extended break (e.g. lunch) at the end of the 4-hour work period in a warm location.

Adapted from ACGIH 2012 TLVs

Amount of Clothing Needed in Cold Environments

$$I = 13.3(T_s - T_a)/M$$

I = clothing insulation needed

(Higher I = more clothing needed; 0 = nude.)

T_s = skin surface temperature in °C

T_a = air temperature in °C

M = metabolic rate in kcal/hr

Convection

$$C = 1.0V^{0.6}(T_a - T_s)$$

C = convection heat transfer in kcal/hr

V = airspeed in m/min

T_a = air temperature in °C

T_s = skin temperature in °C

If air temperature is higher than skin temperature, heat is added to the metabolic load (+C).
If skin temperature is higher than air temperature, heat is lost from the metabolic load (–C).
This assumes a skin area of 1.8 m².
This assumes a skin temperature of 35°C for heat stress conditions.

Convection, Radiation, and Evaporation

The effects of convection, radiation, and evaporation may be calculated using a series of related calculations. All these formulas feed into the heat stress index formula to determine the heat loading on an individual.

Evaporation

$$E_{max} = 2.0V^{0.6}(P_{ws} - P_{wa})$$

E_{max} = maximum evaporative heat loss in kcal/hour

V = airspeed in m/min

P_{ws} = vapor pressure of water at skin temperature in mmHg

P_{wa} = vapor pressure of air temperature in mmHg

(Use psychometric chart to find vapor pressure.)

Evaporation is a cooling mechanism, so E is always negative (–E).

Heat

Working in hot conditions requires specific safeguards.

One of the easiest and most effective methods to combat heat-related disorders is to alternate periods of work with periods of rest. The frequency and duration of rest periods will vary depending on the type of work.

Heat Calculations

$M \pm C \pm R - E = 0$

metabolic heat ± convection ± radiation – evaporation = 0

metabolic costs = calorie consumption
kcal/hour = calories/hour
Use metabolic charts to determine kcal/hour from activity.

Heat Exposure TLV®—WBGT in °C (°F)

Type of Work/Rest

Each Hour	Light	Moderate	Heavy
Continuous	30.0°C (86°F)	26.7°C (80°F)	25.0°C (77°F)
75/25 minutes	30.6°C (87°F)	28.0°C (82°F)	25.9°C (78°F)
50/50 minutes	31.4°C (89°F)	29.4°C (85°F)	27.9°C (82°F)
25/75 minutes	32.2°C (90°F)	31.1°C (88°F)	30.0°C (86°F)

Reference: ACGIH 2012 TLVs®

Heat Stress Calculation

WBGT = 0.7WB + 0.3GT indoors with no solar heat load
WBGT = wet bulb globe temperature
  WB = wet bulb
  GT = globe temperature

WBGT = 0.7WB + 0.2GT + 0.1DB outdoors with solar load
  WB = wet bulb
  GT = globe temperature
  DB = dry bulb

Heat Stress Index

$HSI = E_{req} \times (100/E_{max})$
HSI = heat stress index

E_{req} = evaporative heat loss required in kcal/hr
E_{max} = maximum evaporative heat loss in kcal/hr
Find: E_{req} from $M \pm C \pm R - E = 0$;
C from convection formula;
R from radiation formula;
E_{max} from evaporation formula.[4]

Heat Illness

Exposure to heat can cause illness and death. The most serious heat illness is heatstroke. Other heat illnesses, such as heat exhaustion, heat cramps, and heat rash, should also be avoided.

There are precautions your employer should take any time temperatures are high and the job involves physical work.

Risk Factors for Heat Illness
- High temperature and humidity, direct sun exposure, no breeze or wind
- Low liquid intake
- Heavy physical labor
- Waterproof clothing
- No recent exposure to hot workplaces

Symptoms of Heat Exhaustion
- Headache, dizziness, or fainting
- Weakness and wet skin
- Irritability or confusion
- Thirst, nausea, or vomiting

Symptoms of Heatstroke
- May be confused or unable to think clearly, pass out, collapse, or have seizures (fits)
- May stop sweating

To Prevent Heat Illness, Your Employer Should
- Provide training about the hazards leading to heat stress and how to prevent them.
- Provide a lot of cool water to workers close to the work area. At least 1 pint of water per hour is needed.
- Schedule frequent rest periods with water breaks in shaded or air conditioned areas.

- Routinely check workers who are at risk of heat stress due to protective clothing and high temperatures.
- Consider supplying protective clothing that provides cooling.

VENTILATION FAN LAWS

Fans, spray hoods, and other ventilation methods remove unwanted particles, gas, or fumes from the working area. Pressure, flows, and draw must all be in the correct balance in order to capture the contaminant and deliver it to a filtering media. When proceeding with calculations, remember to use consistent units.

Contaminant Capture

Collectors, precipitators, and filters are used to capture air contaminants. Cyclone collectors are generally inexpensive, have a low pressure drop, are low maintenance, and act on the principle of centrifugal action. Electrostatic precipitators are best for capturing minute particles but are rather expensive. Wet collectors such as spray chambers, packed towers, and venturi-type collectors have varying success with different contaminants and varying pressure drops.

Condition of Dispersion	Examples	Capture Velocity (FPM)
Released with practically no velocity into quiet air	Evaporation from tanks; degreasing	50–100
Released at low velocity into moderately still air	Spray booths; intermittent container filling; low-speed conveyor transfers; welding; plating; pickling	100–200
Active generation into zone of rapidly moving air	Spray painting in shallow booths; barrel filling; conveyor loading; crushers	200–500
Released at high initial velocity into zone of very rapidly moving air	Grinding; abrasive blasting; tumbling	500–2000

Capture velocity and exhaust velocity are related to duct velocity. The 10% Rule states that as one moves 1 diameter out from the face, the capture velocity is ±10% of duct velocity. The 30% Rule states that as one moves 1 diameter from the exit, the exhaust velocity is ±30% of duct velocity.

Duct Balancing

Balancing ducts is often a delicate operation. The combining of ducts makes for interesting results. The flow depends on the air velocity and the cross-sectional area of the duct.

$$Q = V \times A$$

flow = velocity × area

Q = flow

V = velocity

A = area

After a diameter change, either constriction or expansion, in a duct, the flow in the duct is the same on either side.

$$V_1 A_1 = V_2 A_2$$

V = velocity

A = area

In a balanced duct, where two ducts combine to form a single duct, the cross-sectional area of the resulting duct is the combination of the cross-sectional area of the two initial ducts.

$$A_t = 1.20(A_1 + A_2)$$

total area = 1.20 × (area of duct$_1$ + area of duct$_2$)

Duct Flow

Flow, static pressure, and horsepower all have a relationship to the revolutions per minute (RPM) of a fan. Flow has a direct relationship with RPM, as shown in the following formula:

$$Q_2/Q_1 = RPM_2/RPM_1$$

Static pressure (head) has a squared relationship with fan RPM, as shown in the following formula:

$SP_2/SP_1 = (RPM_2/RPM_1)^2$

Horsepower has a cubed relationship with fan RPM, as shown in the following formula:

$Hp_2/Hp_1 = (RPM_2/RPM_1)^3$

Fan static pressure ratio:

$SP_{fan} = SP_{out} - SP_{in} - VP_{in}$

   SP = static pressure

   VP = vapor pressure

Duct flow formula:

$$TP = SP + VP$$

total pressure = static pressure + vapor pressure

      SP = static pressure, pressure against walls of conduit

      VP = vapor pressure

      TP = total pressure

VP is always positive.

VP, SP, TP in inches of water

Duct Flow Velocity and Vapor Pressure

The velocity of the air in a duct is related to the vapor pressure in the duct and also to the duct diameter. As the duct diameter increases, the velocity decreases. To increase the static pressure within a duct, increase the velocity of the air in the duct. Two devices for measuring air velocity are anemometers, which measure high air velocity, and velometers, which measure lower air velocity. The relationship of velocity to vapor pressure is illustrated by:

$$V = 4{,}005\sqrt{VP}$$

velocity = 4,005 × square root of vapor pressure

Solvents and TLVs®

Many solvents used in a spray booth have a threshold limit value (TLV®) associated with their vapor. To keep the vapors below the TLV®, a calculation is needed to establish the flow (in CFM) that is necessary.

$Q = [403 \times$ specific gravity of material $\times 10^6 \times$ pints/hour evaporation rate $\times K$ (safety factor)]/molecular weight of material $\times$ TLV®

When working with a solvent, it is also necessary to take into account the lower explosive limit (LEL) and add dilution air to have the concentration lowered to 25% of LEL. An example is offered from 29 *CFR* 1910.94(c)(6)(ii).

To determine the volume of air in cubic feet necessary to dilute the vapor from 1 gallon of solvent to 25% of the lower explosive limit, apply the following formula:

$$\text{Dilution volume required per gallon of solvent} = 4(100 - \text{LEL})$$
$$\text{(cubic feet of vapor per gallon)/LEL}$$

Using toluene as the solvent:
1. LEL of toluene from the following table, column 2, is 1.4 percent.
2. Cubic feet of vapor per gallon from the following table, column 1, is 30.4 cubic feet per gallon.
3. Dilution volume required = $4(100 - 1.4)(30.4)/1.4 = 8,564$ cubic feet.
4. To convert to cubic feet per minute of required ventilation, multiply the dilution volume required per gallon of solvent by the number of gallons of solvent evaporated per minute.

Lower Explosive Limit of Some Commonly Used Solvents

Solvent	Cubic ft per Gallon of Vapor of Liquid at 70°F	Lower Explosive Limit in % by Volume of Air at 70°F
Acetone	44.0	2.6
Amyl Acetate (iso)	21.6	1.01
Amyl Alcohol (n)	29.6	1.2
Amyl Alcohol (iso)	29.6	1.2
Benzene	36.8	1.41

Butyl Acetate (n)	24.8	1.7
Butyl Alcohol	35.2	1.4
Butyl Cellosolve	24.8	1.1
Cellosolve	33.6	1.8
Cellosolve Acetate	23.2	1.7
Cyclohexanone	31.2	1.11
1,1 Dichloroethylene	42.4	5.9
1,2 Dichloroethylene	42.4	9.7
Ethyl Acetate	32.8	2.5
Ethyl Alcohol	55.2	4.3
Ethyl Lactate	28.0	1.51
Methyl Acetate	40.0	3.1
Methyl Alcohol	80.8	7.3
Methyl Cellosolve	40.8	2.5
Methyl Ethyl Ketone	36.0	1.8
Methyl n-Propyl Ketone	30.4	1.5
Naphtha (VM&P) (76° Naphtha)	22.4	0.9
Naphtha (100° flash) Safety Solvent— Stoddard Solvent	23.2	1.0
Propyl Acetate (n)	27.2	2.8
Propyl Acetate (iso)	28.0	1.1
Propyl Alcohol (n)	44.8	2.1
Propyl Alcohol (iso)	44.0	2.0
Toluene	30.4	1.4
Turpentine	20.8	0.8
Xylene (o)	26.4	1.0

1 At 212° F.

Occupational medicine encompasses many safety specialties. Exposure to toxic agents may cause illness, injury, or disease. Repetitive motion may cause ergonomic aggravations. Often, a worker's complaint first falls on the ears of the health care professional. This member of the health and safety team must be aware of the unique circumstances of occupational situations to accurately diagnose and treat potential health problems not usually found in the general population. Companies fortunate to have their own health care professional may become immersed in their industry's specific health hazards. Companies who must rely on a medical service or clinic should engage these professionals in a continuing dialogue to better protect their workers.

COMMON DISEASES AND THEIR CAUSATIVE AGENTS

- Anthracosis—a lung condition caused by exposure to coal dust; also known as black lung
- Anthrax—a highly virulent bacterial infection contracted from animals and animal products such as wool
- Bagassosis—a respiratory disorder caused by breathing fungi found in sugar cane dust
- Brucellosis—a group of infectious diseases caused by an organism of the *Brucella* genus; one source is unpasteurized milk from cows suffering from Bang's disease (infectious abortion)
- Byssinosis—a pulmonary disease caused by prolonged exposure to air concentrated with hefty amounts of cotton or flax dust
- Cataract—opacity in the lens of the eye that may obscure vision; caused by, for example, exposure to infrared (IR) light

- Chrome nose, nickel nose—condition resulting from exposure to chromic acid that destroys nasal passages
- Erysipelosis (erysipeloid)—a bacterial skin condition caused by exposure to unprocessed fish or meat
- Erythema—reddening of the skin caused by excessive exposure to sunlight or UV exposure (sunburn)
- Glanders—an ulcerative skin condition caused by exposure to infected horses or mules
- Histoplasmosis—a respiratory condition caused by exposure to bird waste; commonly referred to as pigeon poop disease
- Legionnaires' disease—a pneumonia-like infection of the lungs caused by inhaling bacteria in the genus *Legionella*; often found in stagnant] water and spread through heating and cooling systems
- Leptospirosis—an infectious condition caused by exposure to bacteria found in wild animal or dog urine
- Newcastle's disease—an avian pneumoencephalitis caused by exposure to birds, especially chickens, that can result in eye conjunctivitis
- Pneumoconiosis—a condition resulting when inorganic dust particles lodge in the lungs and cause breathing problems
- Presbycusis—loss of hearing due to age
- Q fever—a rickets-like disease found in livestock handlers
- Sensorineural hearing loss—loss of hearing due to overexposure to excessive noise
- Siderosis—a type of pneumoconiosis (disease of the lungs) caused by inhalation of iron or welding fumes
- Silicosis—a type of pneumoconiosis caused by inhalation of sand or silica dust
- Tetanus—an infectious, acute disease resulting from exposure to bacillus bacteria that causes muscle contractions; also known as lockjaw

- Trichinosis—a gastrointestinal disease caused by ingestion of uncooked pork or bear

COMMON ERGONOMIC SYNDROMES

- Carpal tunnel syndrome—inflammation of the tendons in the wrist that affects the median nerve
- DeQuervain's disease—swelling of the tendon sheath of the thumb
- Epicondylitis—tennis elbow
- Raynaud's syndrome—constriction of the blood vessels in the hand; usually due to vibrations
- Tendinitis—swelling of a tendon
- Tenosynovitis—swelling of the sheath surrounding a tendon
- Trigger finger—a particular type of tenosynovitis in which the tendon becomes nearly locked, which pulls the finger toward the palm with a jerky movement

HEALTH HAZARDS

- Analytic epidemiology—proves or disproves a hypothesis
- Asphyxiant—a chemical agent (usually a gas) that interferes with normal inspiration [13]
  - Simple asphyxiant—not toxic alone but reduces oxygen concentration and replaces O_2 in the blood. Examples: CO_2, N_2, H_2
  - Chemical asphyxiant—reduces body's ability to absorb O_2 in the blood and is toxic. Examples: CO, H_2S, HC (hydrogen cyanide)
- Bacteria—can live without a host; cause injury by producing toxins, producing hypersensitivity, or invading and destroying cells
- Dosage—amount of chemical per unit of body weight

- Dose threshold—minimum dosage that produces the desired effect
- Endemic—localized outbreak of illness
- Epidemic—universal outbreak of illness
- Epidemiology, descriptive—number of cases per time period
- Epidemiology, prevalence—number of cases per total subjects
- Hepatotoxic—toxic to the liver
- Incidence rate—number of new cases per number at risk
- LC_{50}—lethal concentration 50; the concentration of a gas where 50% of the subjects will perish; route of entry through inhalation; usually expressed in ppm
- LD_{50}—lethal dose 50; the dose of a liquid or solid where 50% of the subjects will perish; numerically determined by weight of chemical per weight of animal; routes of entry through ingestion or absorption
- Mutagen—a substance that causes permanent changes to DNA; passed on to future generations through mutations
- Neurotoxin—affects the central nervous system. Examples: chronic mercury exposure, mad hatters' disease
- Pandemic—widespread outbreak of illness; routes of entry are inhalation, ingestion, absorption, injection
- Sensitizers—substances to which repeated exposure can result in severe allergic reaction. Examples: formaldehyde, isocyanates
- Teratogen—substance such as thalidomide that causes irreversible damage to the fetus
- Virus—needs host to survive and reproduce

Dose = Quantity of Chemical × Exposure Time

Biological Response
- *Independent effect*—Substances exert their toxicity independently of each other.
- *Additive effect*—The combined effect of exposure to two chemicals that have both the same mechanism of action and the same target organ; is equal to the sum of the effects of exposure to each chemical when given alone (e.g., 3 + 5 = 8). In the absence of any data to the contrary, chemicals are assumed to interact in an additive manner. Examples: two different organophosphate insecticides and inhibition of acetyl cholinesterase at neuromuscular junctions
- *Synergistic effect*—The combined effect of exposure to two chemicals is much greater than the sum of the effects of each substance when given alone (e.g., 3 + 5 = 30); each substance magnifies the toxicity of the other.
- *Potentiating effect*—One substance, having very low or no significant toxicity, enhances the toxicity of another substance (e.g., 0 + 5 = 15); the result is a more severe toxicity than what each toxic substance would have produced by itself.
- *Antagonistic effect*—An effect where two chemicals interfere with each other's toxic actions (e.g., 4 + 6 = 8) or where one chemical interferes with the toxic action of the other chemical, such as in antidotal therapy (e.g., 0 + 4 = 2).

Categories of Hazards in Industrial Toxicology

Class of Hazard	Affected Organs	Possible Signs and Symptoms	Selected Examples
Irritants	Exposed tissues (e.g., mucous tissues)	Pain, fluid accumulation	Acids and acid vapors, sulfur dioxide, ozone, ammonia, chlorine
Simple Asphyxiants	Body cells (O_2 is blocked from the lungs)	Confusion, collapse, unconsciousness	Nitrogen, argon, helium, carbon dioxide, methane (natural gas)
Chemical Asphyxiants	Body cells (interfere with oxygenation of body cells)	Confusion, headache, collapse, unconsciousness	Carbon monoxide, hydrogen sulfide, inorganic cyanides
Anesthetics or Central Nervous System Depressants	Central nervous system	Dizziness, drowsiness, collapse, unconsciousness	Liquids and vapors of many organic solvents (e.g., alcohols, ethers, esters, chlorinated hydrocarbons, toluene, xylene, benzene)
Agents that can cause lung illnesses (e.g., fluid accumulation, tissue scarring, cancer)	Lungs, lining of either the lungs or the abdomen	Breathlessness, chest pain, cough, weakness	Asbestos, crystalline silica (e.g., quartz), coal, radioactive substances, beryllium and its compounds, welding fumes, cotton fibers, arsenic and its compounds, hydrogen fluoride, phosgene, nitrogen dioxide

Carcinogens (agents that are known or suspected of causing cancer in humans)	Various organs, depending on the agent	Pain, coughing, tumors, various other symptoms	Confirmed carcinogens include: asbestos, radioactive substances, polycyclic aromatic hydrocarbons (PAH), arsenic and its compounds, benzene, hexavalent chromium compounds, coal tars and pitches, vinyl chloride Suspected carcinogens include: acrylonitrile, benzidine-based dyes, benzo[a]pyrene, beryllium and its compounds, cadmium and its compounds, carbon tetrachloride, creosote, ethylene oxide, formaldehyde gas, 2,3,7,8-tetrachloro-dibenzo[p]dioxin (TCDD), trichloroethylene, tetrachloroethylene, and crystalline silica dust
Nephrotoxins	Kidneys	Symptoms vary for different agents	Most heavy metals and their compounds (e.g., lead, mercury, chromium, uranium), some halogenated hydrocarbons (e.g., trichloroethylene, chloroform, carbon tetrachloride), 2,4,5-trichlorophenoxyacetic acid (2,4,5-T), polychlorinated biphenyls (PCBs)
Hepatotoxins	Liver	Symptoms vary for different agents	Some halogenated hydrocarbons (e.g., carbon tetrachloride, chloroform, trichloroethylene, tetrachloroethylene), ethyl alcohol, allyl alcohol, urethane monomer, hydrazine, cerium and its compounds, beryllium and its compounds, some pharmaceuticals

Chemical Allergens (sensitizers)	Various tissues, especially skin and eyes	Itching, swelling, inflammation	Formaldehyde, beryllium and its compounds, toluene-2,4-diisocyanate (TDI), creosote, some acrylates, epoxy resins and components, coal tar and its derivatives, some organic dyes, turpentine, some woods, poison ivy and oak, white sumac, some pharmaceuticals
Genotoxic and Fetotoxic Agents	Genotoxic: affects the genetic material of reproductive cells Fetotoxic: affects a fetus	Genotoxins: may be few immediate signs Fetotoxins: deformity or loss of a fetus	Benzene, toluene, xylene, ethyl alcohol, carbon disulfide, carbon monoxide, lead and its compounds, mercury and its compounds, arsenic and its compounds, cadmium and its compounds, radiation and radioactive substances, chlorinated phenoxyacetic acids, paraquat, diquat, PCBs, TCDD, ethylene oxide, dinitrobenzene, some pharmaceuticals

Knowledge of basic math and science principles as applied to safety, health, and environmental management is necessary for the safety professional. Principles, laws, and formulas provide the mathematical basis for calculations. Completing accident reconstructions, industrial hygiene calculations, stress calculations, and many other computations requires a solid foundation in math. Newer calculators have the capability to do much of the more mundane work, but they cannot replace the human element.

Equivalents, constants, and conversions allow calculations to be performed in a straightforward manner. A good calculator will provide many of the conversions from English to metric units, but a familiarity with some of the more important conversions is necessary.

CONVERSIONS:

- 3.758 liters = 1 gallon
- 2.2 lbs = 1 kilogram
- 2.54 cm = 1 inch
- 0.433 psi/ft = per foot of rise
- 1.47 feet per second = 1 mph
- 9.8 kg/cubic meter (m^3) (water)
- 5,280 feet = 1 mile
- 1 foot = 0.3048 cubic meter (m^3)
- 1 cu ft = 7.48 gallons
- 8.34 lbs = 1 gallon of water
- 1 cm^3 = 1 gram
- weight of 1 cm^3 of water = 1 gram
- 1 lb = 454 g
- 16 oz = 454 g
- 1 lb = 4.4 newtons
- 1% by volume = 10,000 ppm

- 1 liter of water = 1,000 grams
- 1 liter = 1,000 milliliters
- 1 meter = 1,000 milliliters
- 1 meter = 100 centimeters
- 1 liter = 1,000 cm^3
- 1 meter = 1,000 millimeters
- 1,000 meters = 1 kilometer

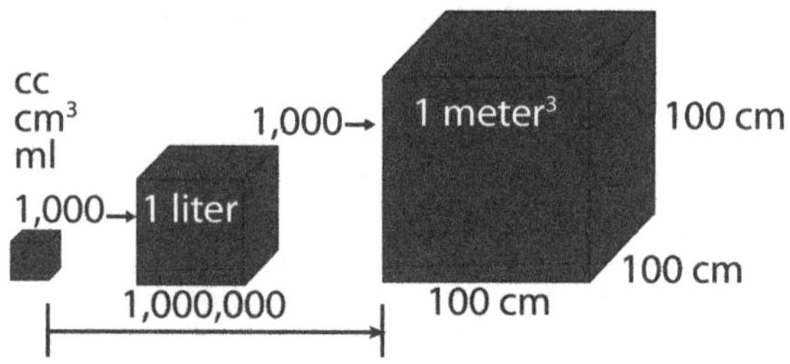

Kilo = 1,000 or 10^3	Deci = 0.1 or 10^{-1} or one-tenth
Mega = 1,000,000 or 10^6	Centi = 0.01 or 10^{-2} or one-hundredth
Giga = 1,000,000,000 or 10^9	Milli = 0.001 or 10^{-3} or one-thousandth
	Micro = 0.000001 or 10^{-6}
	Nano = 0.000000001 or 10^{-9}

AIR

- 1 atm = 14.7 psi = 29.92 inches of mercury (Hg) = 101,325 pascals = 760 torr = 760 mm Hg
- psig—The actual reading on the gauge
- psia—The gauge reading plus 14.7 psi

Universal Gas Constant (R) =
- 0.7302 ft^3 atm/lb mole °R
- 10.73 ft^3 psia/lb mole °R
- 62.36 liters mmHg/mole °K

- 8.32 × 107 ergs/°C gram mole
- 8.314 m^3 Pa/mole °K
- 0.08314 liter bar/mole °K
- 1.987 cal/mole °K
- 1.987 BTU/lb mole °R

AVOGADRO'S NUMBER AND MOLES

- 6.024×10^{23} gm mole
- 6.024×10^{23} molecules = 1 mole
- 22.4 l/gm mole at 0°C, 1 atm
- 24.45 l/gm mole at 25°C, 1 atm

DISTANCE, LENGTH, AREA

- 1 mile = 5,280 feet
- 1 inch = 25.4 mm
- 1 foot = 30.48 centimeters = 0.3048 meter
- 1 acre = 0.4047 hectare = 43,560 square feet
- A circle = 360 degrees = 2π radians

ELECTRICITY AND MAGNETISM

- 1 tesla = 10,000 Gauss (magnetism)
- capacitance (faraday) = 9.65×10^4 coulombs
- 1 coulomb = 6.25×10^{18} electrons
- voltage = current (amperage) × resistance
- power (wattage) = voltage × current
- 1 ampere = 1 coulomb/sec
- 1 amphour = 3,600 coulombs
- 1 watt = 1 joule/sec = 1 ampvolt = 1.341×10^{-3} hp

Forces
- Acceleration due to gravity = 32.2 ft/sec^2 = 9.8 m/sec^2
- 1 newton = 10^5 dynes = 0.22481 lbf
  - lbf = pound force = weight
- 1 ft lb = 32.174 (lbm ft)/sec^2 = 4.4482 newtons
  - lbm = mass

GREEK ALPHABET

A	α	alpha	I	ι	Iota	P	ρ	rho
B	β	beta	K	κ	kappa	Σ	σ	sigma
Γ	γ	gamma	Λ	λ	lambda	T	τ	tau
Δ	δ	delta	M	μ	mu	Y	υ	upsilon
E	ε	epsilon	N	ν	nu	Φ	φ	phi
Z	ζ	zeta	Ξ	ξ	xi	X	χ	chi
H	η	eta	O	o	omicron	Ψ	ψ	psi
Θ	θ	theta	Π	π	pi	Ω	ω	omega

HEAT AND ENERGY

- 1 BTU (British thermal unit) = 1,054.8 joules
- 1 joule (work, heat, energy) = 10^7 ergs = 0.239 calorie
- 1 gram calorie = 4.19 joules = 4.19×10^7 ergs
- 1 horsepower = 746 watts, 550 ft lb/sec
- 1 watt = 1 joule/sec, 1 ampvolt

LIGHT AND SOUND

- frequency (hertz) = cycles/sec
- Planck's constant = 6.626×10^{-27} erg sec
- 1 candela/ft^2 = 3.38×10^{-3} lambert
- 1 candlepower = 12.57 lumens
- 1 foot-candle = 10.76 candela/m^2 of air
- 1 lumen = 1 candela
- 1 lumen/ft^2 = 1 foot-lambert = 1 foot-candle
- 1 lumen/m^2 = 1 lux
- 1 candle/in.^{2} = 452.4 foot-lamberts = 0.487 lambert
- speed of light in a vacuum = 300,000 km/sec
  = 186,000 miles/sec
- speed of sound at 0°C, sea level = 331.5 m/sec
  = 1,087 ft/sec = 741 miles/hour (Mach 1)

- speed of sound at 60°C, sea level = 340.3 m/sec
    = 1,116 ft/sec
    = 761 miles/hour
- speed of sound at 60°C, in water = 1,450 m/sec
    = 4,750 ft/sec
    = 3,240 miles/hour

MOHS SCALE OF HARDNESS

Mohs Hardness Scale		Approximate Hardness of Common Objects
Talc	1	
Gypsum	2	Fingernail (2.5)
Calcite	3	Copper penny (3.5)
Fluorite	4	Iron nail (4.5)
Apatite	5	Glass (5.5)
Feldspar	6	Steel file (6.5)
Quartz	7	Streak plate (7.0)
Topaz	8	
Corundum	9	
Diamond	10	

Prefixes

Tetra/terra	1,000,000,000,000	10^{12}	T
Giga	1,000,000,000	10^{9}	G
Mega	1,000,000	10^{6}	M
Kilo	1,000	10^{3}	k
Hecto	100	10^{2}	h
Deka	10	10^{1}	da
Deci	0.1	10^{-1}	d
Centi	0.01	10^{-2}	c
Milli	0.001	10^{-3}	m
Micro	0.000001	10^{-6}	μ
Nano	0.0000000001	10^{-9}	n
Pico	0.0000000000001	10^{-12}	p

An Angstrom unit is 10^{-10} meter.

RADIATION

- 1 curie = 3.7×10^{10} disintegrations per second
      = 3.7×10^{10} becquerels
- 1 becquerel = 2.7027×10^{-11} curie
- 1 rad = 10^{-2} gray
- 1 rem = 10^{-2} sievert
- 1 roentgen = 2.58×10^{-4} coulomb/kg of air

SPECIFIC GRAVITY

- density of object/density of water

TEMPERATURE

- Celsius (C) = (Fahrenheit − 32)/1.8
- Kelvin (K) = Celsius + 273.16
- Rankine (R) = Fahrenheit + 459.69

WATER

- 1 gallon = 8.34 lb = 231 in.^{3} = 3.7854 liters = 3785.4 cc
- 1 ft^3 = 62.4 lb (density) = 0.433 psi (62.4/144) = 7.48 gal
- 1 gm/cm^3 = 1.94 slugs/ft^3

BASIC MOLECULAR TERMS

The following terms describe the relationships within the atom and involved in chemical reactions:
- Atomic mass = protons + neutrons
- Atomic mass − atomic number = number of neutrons
- Atomic number = number of protons
- Avogadro's number = the number of molecules in 1 mole, or the gram molecular weight; 6.024×10^{23} molecules

- Deflagration = rapidly burning fire; flame speed slower than speed of sound
- Detonation = rapidly burning fire; flame speed greater than speed of sound
- Endothermic = consumes energy
- Exothermic = produces energy
- Hypergolic = violent reaction
- Isomer = same number of atoms but arranged differently
- Isotope = same atomic number, but different atomic mass (protons the same but neutrons different)
- Mole = mass/molecular weight
- Molecular weight of air = approximately 30
- Molecular weight of a substance = the total molecular weight of all the elements
- Number of protons = number of electrons, unless isotope or ion
- Oxidation = loss of electrons
- Percent solution = moles/density
- Pyrolysis = decomposes in presence of heat
- Pyrophoric = ignites spontaneously in air
- Reduction = gain of electrons (gains negativity)

Common Chemical Formulas

Alcohol group	OH
Benzene	C_6H_6
Carbon tetrachloride	CCl_4
Ethane	C_2H_6
Ethyl group	C_2H_5
Methane	CH_4
Methanol	CH_3OH
Methyl group	CH_3
Propane	C_3H_8
Sodium hydroxide	NaOH
Sodium nitrate	$NaNO_3$

GAS AND AIR CONCEPTS

The relationship between molecules in a gas—or the effects of volume, temperature, pressure, and elevation—is expressed in several gas laws:
- Absolute pressure = gauge air pressure + 14.7

Boyle's Law
- $P_1V_1 = P_2V_2$ (constant temperature)
  - P_1 = pressure of gas$_1$
  - V_1 = volume of gas$_1$
  - P_2 = pressure of gas$_2$
  - V_2 = volume of gas$_2$

Charles' Law
- $V_1/T_1 = V_2/T_2$ (constant pressure)
  - V_1 = volume of gas$_1$
  - T_1 = absolute temperature of gas$_1$
  - V_2 = volume of gas$_2$
  - T_2 = absolute temperature of gas$_2$

Combined Gas Law
- $P_1V_1/T_1 = P_2V_2/T_2$
  - P_1 = pressure of gas$_1$
  - V_1 = volume of gas$_1$
  - T_1 = absolute temperature of gas$_1$
  - P_2 = pressure of gas$_2$
  - V_2 = volume of gas$_2$
  - T_2 = absolute temperature of gas$_2$

Conversion of TLVs® in ppm to mg/m³

TLVs® for gases and vapors are usually established in terms of parts per million of substances in air by volume (ppm). For convenience to the user, these TLVs® are also listed with molecular weights. Where 24.45 = molar volume of air in liters at normal temperature and pressure (NTP) conditions (25°C and 760 torr), the conversion equation for mg/m³ is:

$$\text{TLV}^{\circledR}\text{ in mg/m}^3 = \frac{(\text{TLV}^{\circledR}\text{ in ppm})(\text{gram molecular weight of substance})}{24.45}$$

Conversely, the equation for converting TLVs® in mg/m³ to ppm is:

$$\text{TLV}^{\circledR}\text{ in ppm} = \frac{(\text{TLV}^{\circledR}\text{ in mg/m}^3)(24.45)}{\text{gram molecular weight of substance}}$$

The above equation may be used to convert TLVs® to any degree of precision desired. When converting TLVs® to mg/m³ for other temperatures and pressures, the reference TLVs® should be used as a starting point. When converting values expressed as an element (e.g., as Fe or as Ni), the molecular value of the element should be used, not that of the entire compound.

Dalton's Law
- Total pressure exerted by a gas mixture equals the total of the pressures of the individual gases

Dalton's Law of Partial Pressure
P_{gas} = pressure of air at a specific altitude × % gas

Gas Constant
- Gas constant (R) = 0.7302 ft³ atm/lb mole °R (Rankine)
  - 10.73 ft³ psia/lb mole °R
  - 62.36 liter mmHg/mole °K
  - 8.32×10^7 ergs/°C × gram mole
  - 8.314 m³ Pa/mole °K
  - 0.08314 liter bar/mole °K
  - 1.987 cal/mole °K
  - 1.987 BTU/lb mole °R

Normal Temperature and Pressure
- NTP = normal temperature and pressure measured at 25°C and 1 atm
- NTP mole of gas at 25°C and 1 atm = 24.45 liters

Standard Temperature and Pressure
- STP = standard temperature and pressure
- STP mole of gas at 0°C and 1 atm = 22.4 liters
- 1 gram mole = 22.4 liters or 22.4 liters/gram mole
- 1 lb mole = 359 ft³ or 1 lb mole will evaporate and fill 359 ft³
- density of air = 0.075 lb/ft³ at STP

Raoult's Law
- Vapor pressure of a volatile solvent in dilute solution is proportional to its mole fraction; as vapor pressure declines, the boiling point increases

Universal Gas Law
- PV = nRT
  - pressure × volume = moles × gas constant × absolute temperature (°R or °K)

Ventilation Standard
- Ventilation standard = 70°F and 1 atm
- 1 atmosphere (atm) =
  - 14.7 psi
  - 760 mmHg
  - 29.92 in. Hg
  - 33.9 ft water
  - 407 in. water
  - 760 torr
  - 101.3 kilopascal
  - 10.35 m water

Volume of Substance
- Volume of substance = amount of substance × moles × STP at specific temperature
  - Example: 100 grams of nitrogen × 1 mole/28 grams × 24.45 liters/mole

pH
- A laboratory hint is to add an acid to water, rather than add water to an acid, as there is less splattering. The mnemonic device is "Always Add Acid"—AAA.
- pH scale = logarithmic scale of the number of

hydrogen (H) ions in solution (per liter) in reciprocal powers of 10
- High pH = large negative power of 10 (weaker solution of H ions); alkaline or base
- Low pH = small negative power of 10 (stronger solution of H ions); acidic or acid

pH of Common Substances

Acetic acid	pH 2.4 (1N)*
Ammonia	pH 11.6 (1N)
Boric acid	pH 5.3 (0.1N)
Distilled water	pH 7.0
Formic acid	pH 2.3 (0.1N)
Hydrochloric acid	pH 0 (0.1N)
Lime	pH 12.4 (saturated)
Sodium hydroxide	pH 14 (1N)
Sulfuric acid	pH 0.3 (1N)
Trisodium phosphate	pH 12.0 (0.1N)
Vinegar	pH 4.0

*Normal

TRIGONOMIC AND GEOMETRIC FORMULAS

Trigonometry is the basis of much of the mathematical aspect of safety. For example, relationships among the different parts of the triangle are used in force equations. Volume, area, and perimeter equations are also used constantly.

BASIC RELATIONSHIPS

- 180 degrees = π radians
- 360 degrees = 2π radians
- 1 radian = $180/\pi$ degrees
- 1 degree = $\pi/180$ radians
- 30 degrees = $30 \times \pi/180$ radians = $\pi/6$ radians
- Speed of a wheel = RPM $\times$ circumference

GEOMETRIC FORMULAS

Triangle
- (a and b = lengths of sides; c = hypotenuse)

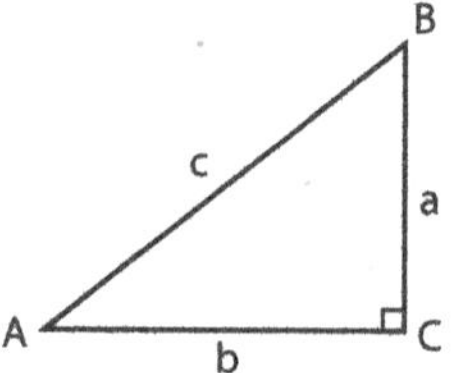

- Primary functions:
  - sine A = a/c
  - cosine A = b/c
  - tangent A = a/b
- Pythagorean theorem: $c^2 = a^2 + b^2$
- Area of a triangle = ½bh

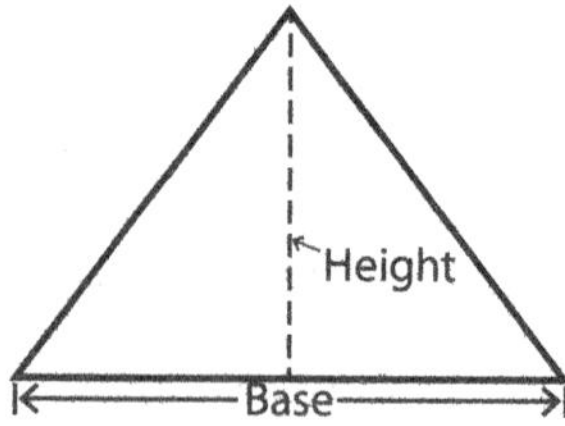

Circle
- Circumference = 2πr or πd
- Area = πr^2
  - r = radius

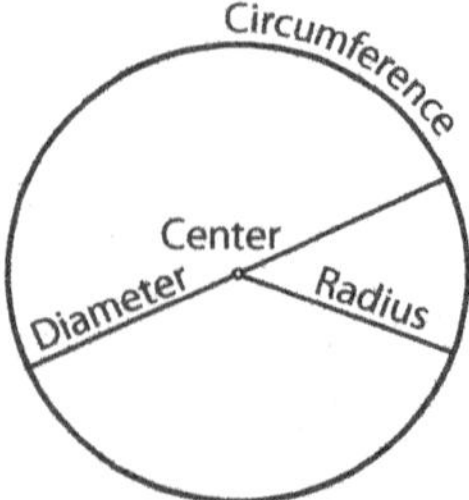

Cone

- Area = π × r × length of side + base area
= p × r × ($\sqrt{\pi r^2 \times h^2}$)
    r = radius
    h = height
Volume = $\pi r^2 h / 3$

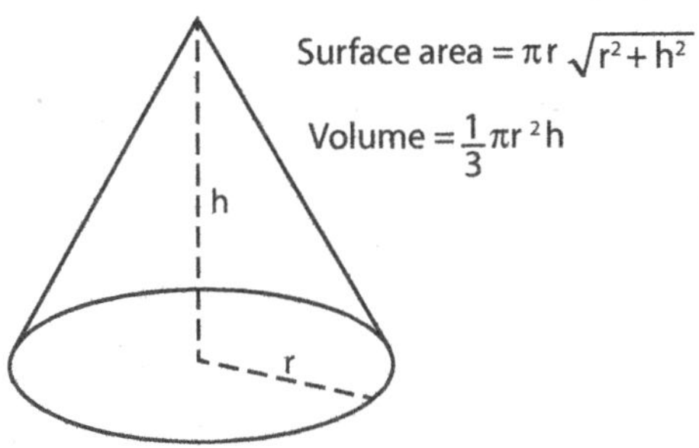

Cube

- Area = $6s^2$
- Volume = s^3

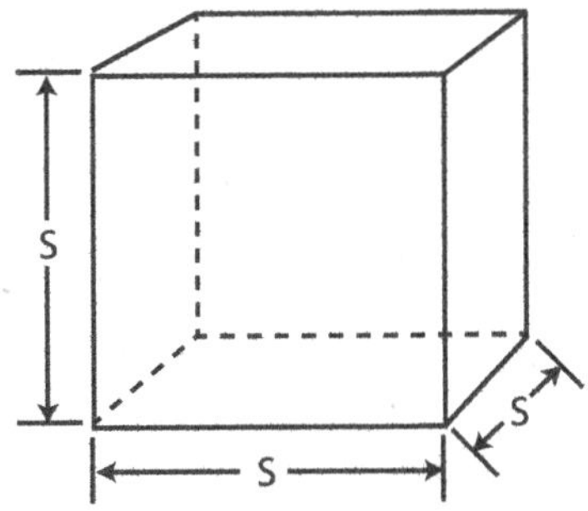

Cylinder

- Total area (including top and bottom)
- = $2\pi rh + 2\pi r^2$
- Volume = $\pi r^2 h$

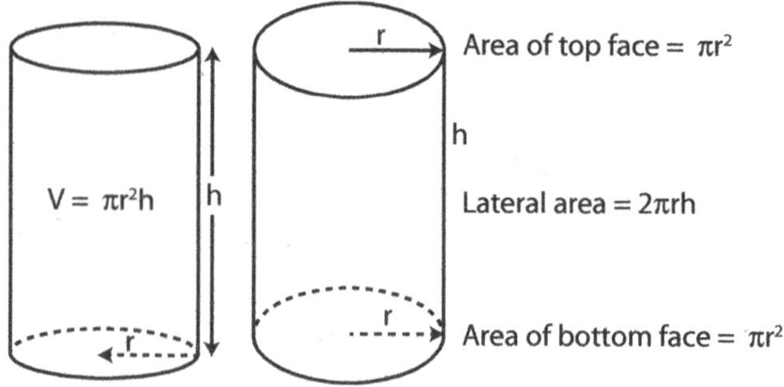

Total surface area

$A = \pi r^2 + \pi r^2 + 2\pi rh$

$A = 2\pi r^2 + 2\pi rh$

$A = 2\pi r(r + h)$

Hexagon

- Area $= (2.5981)s^2$

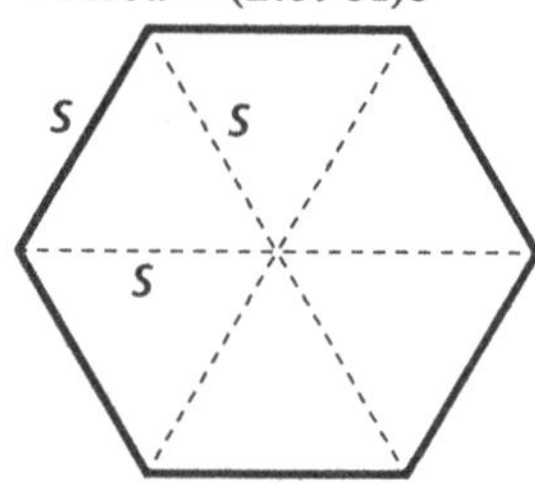

Octagon

- Area $= (4.8284)a^2$

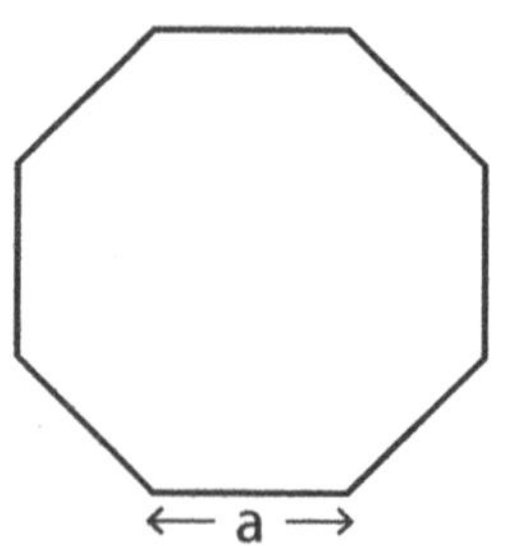

Parallelogram
- Area = bh

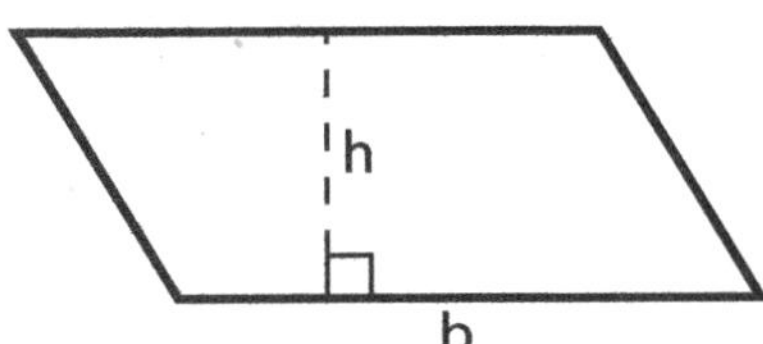

Pentagon
- Area = $1.7205a^2$

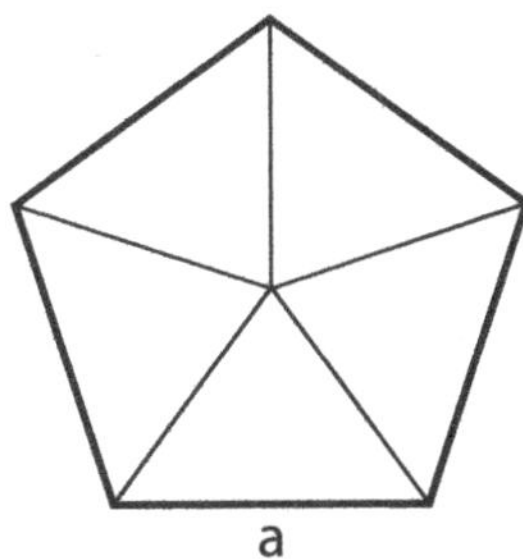

Pyramid
- Volume = ⅓ base area × height

Square-based Pyramid

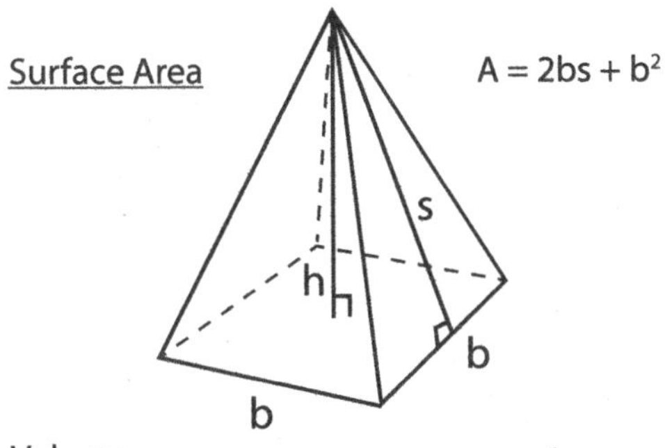

Rectangle

- Area $= l \times w$

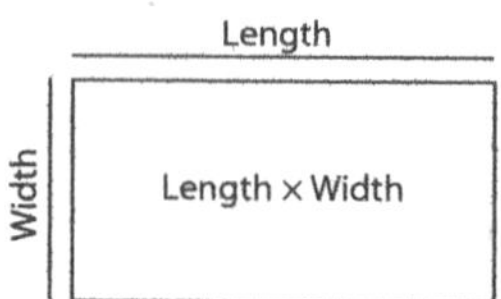

- Volume $= l \times w \times h$

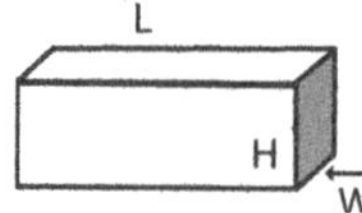

Volume $= L \times W \times H$
Volume $=$ Length $\times$ Width $\times$ Height

Sphere

- Area $= 4\pi r^2$
- Volume $= \frac{4}{3}\pi r^3$

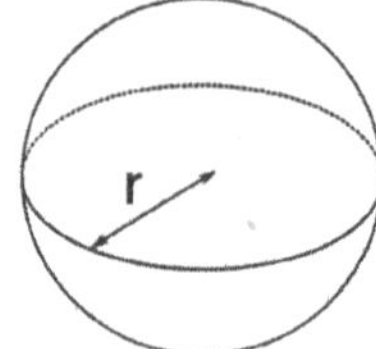

Square

- Area $= s^2$

Trapezoid

- Area $= \frac{1}{2}h(b_1 + b_2)$

Area $= \frac{1}{2}h(b_1 + b_2)$
Area $= \frac{1}{2}(\text{height}) \times (\text{base}_1 + \text{base}_2)$

SAFETY ECONOMICS

As the economic basis of safety becomes more important, so does the ability of the safety professional to adequately describe, in monetary terms, a loss or reduction in risk. Relating the cost of an injury to the number of units of product needed to cover that loss is a concept that is understood by the businesspeople who run companies.

Cost of loss = P × N × U

$\quad$ P = profit margin in % (expressed as decimal)
$\quad$ N = number of products needed to cover loss
$\quad$ U = unit selling price

Cost of loss = P × V

$\quad$ P = profit margin in % (expressed as decimal)
$\quad$ V = $ volume of business

Example:
$\quad$ $100 loss = 5% profit margin (0.05) × volume
$\quad$ $2,000 = volume needed to cover loss

INTEREST

Simple

$i = p \times r \times t$
i = interest
p = principal
r = annual interest rate (decimal)
t = time in years

Compound

$a = P \times [1 + (r/m)]^{m \times t}$
a = value of investment after t years
p = amount of original investment
r = interest rate (decimal)
m = number of compoundings per year
t = years
$m \times t$ = total compoundings

This formula can also be used to find present and future values of money calculations, where $1 + r/m =$ interest accumulation factor over t years.

a = future value

p = present value

INVESTMENT GROWTH CALCULATIONS

Growth Time

Doubling time of an investment = dt

$\qquad = [\ln(2) \times 1/m]/\{\ln[1 + (r/m)]\}$

dt = doubling time

ln(2) = natural log of 2

m = number of compoundings per year

r = interest rate (decimal)

Easier method for doubling: ln(2)/r

Much easier method for doubling: 70/i

i = interest rate (as a whole number, not a decimal)

Tripling time = ln(3)/r

Quadrupling time = ln(4)/r

Future Value of Annual Payments

$FV = A \times \{[(1 + i)^n - 1]/i\}$

FV = future value

A = annuity amount

i = interest rate in percent per period

n = number of payments

Future Value of a Lump Sum

$FV = p(1 + i)^n$

FV = future value

p = present value

i = annual interest rate

n = number of years

Present Value of a Series of Payments

$PV = A \times \{[(1 + i)^n - 1]/i(1 + i)^n\}$

PV = present value

A = annuity amount

i = interest rate in percent per period

n = number of years

Sinking Funds Calculation
(What to invest to reach an amount in a given number of years)

$P = (FV \times i)/[(1 + i)^n - 1]$

P = payment size

FV = future value

i = interest rate per compounding period

n = total number of installment payments

BASIC PHYSICS FORMULAS

$f = m \times a$
force = mass × acceleration

$f = ma/g_c$
force = mass × acceleration/gravitational constant

$KE = (mv^2)/2g_c$
kinetic energy = mass × velocity^2/2 × gravitational constant

$p = fv$
power = force × velocity

$PE = mgh/g_c$
potential energy = mass × gravity × height/gravitational constant =
weight × height

$\rho = mv$
momentum = mass × velocity

$w = fs$
work = force × distance

$w = mg$
weight = mass × gravity

$w = ma$
weight = mass × acceleration

COEFFICIENT OF FRICTION

The coefficient of friction is the measure of the slipperiness of a surface. It is symbolized by the Greek letter μ and is unitless. The coefficient of friction of carpet is different from that of glass. The coefficient is used in formulas where sliding or slipping is a key component. A common formula is:

$$f = \mu \times N$$

force = coefficient of friction × weight

A common use of the coefficient of friction is in the sliding box problem. In this problem, a box is placed on an incline. Based on the weight of the box, the angle of incline, and the coefficient of friction of the incline, the goal is to determine whether the box will slide down the incline or stick to the incline. In general, the force down the length of a ramp is a sine force. The normal weight of a box is a cosine function (normal equals weight of box perpendicular to the ground). The down force minus the normal force indicates whether the box will stick or slide. If it is a positive value, the box will not slide. If it is a negative value, the box will slide (force of the slide is greater than the force needed to overcome friction).

$$f_1 d_1 = f_2 d_2$$

$\text{force}_1 \times \text{distance}_1 = \text{force}_2 \times \text{distance}_2$

MOTOR VEHICLE ACCIDENT FORMULAS

Motor vehicle accidents (and similar accidents) spawn another series of related formulas.

To find whether a load will tip over while going around a corner, use the following equation:

$$V_{max} = \sqrt{g \times r \times (d_{cg}/2h_{cg})}$$

V_{max} = max velocity before tipping

   g = gravitational constant = 32.2 in English units, 1.0 in metric units

   r = radius of turn

  d_{cg} = distance from center of gravity to wheel (center line of vehicle to wheel), in ft

  h_{cg} = height of center of gravity from the ground, in ft

To find the velocity of a vehicle in locked wheel braking, the following formulas are useful:

$$V_{mph} = \sqrt{30 \times s \times \mu \times n}$$

v = initial velocity in mph

s = distance (length of skid marks)

μ = coefficient of friction for tires on pavement

n = 1 for 4-wheel skid, 0.75 for 3 wheels, 0.50 for 2 wheels

Or similarly:

$$s = v^2/30 \times \mu$$

s = distance

$v = \text{velocity} = \sqrt{(30 \times KE)/\text{weight}}$

μ = coefficient of friction

If the coefficient of friction of the tires on the pavement is known, then it can be determined whether the vehicle will skid before tipping by comparing the frictional force and the centrifugal force.

Frictional force (FF) = $\mu \times N$

Centrifugal force (CF) = mv^2/r

μ = coefficient of friction

N = normal force (weight)

m = mass (weight/g_c)

g_c = gravitational constant = 32.2 ft/sec^2

v = maximum velocity before tipping

r = radius of turn

If FF > CF, then vehicle will tip.

If CF > FF, then vehicle will skid before tipping.

Energy sources—including electrical, mechanical, hydraulic, pneumatic, chemical, thermal, and other types—in machines and equipment can be hazardous to workers. During the servicing and maintenance of machines and equipment, the unexpected startup or release of stored energy from equipment could cause injury to employees.

ENERGY ISOLATION—LOCKOUT/TAGOUT (LOTO)

The failure to control hazardous energy accounts for nearly 10% of the serious accidents in many industries. Proper lockout/tagout (LOTO) practices and procedures safeguard workers from the release of hazardous energy by disabling machinery or equipment. Implementing effective measures for controlling different types of hazardous energy is critical.

- Employers need to have proper lockout/tagout (LOTO) practices and procedures that include specific actions to address and control hazardous energy during servicing and maintenance of machines and equipment.
- Employers are also required to train each worker to ensure that he or she knows, understands, and is able to follow the applicable provisions of the hazardous energy control procedures. Workers must be trained in the purpose and function of the energy control program and have the knowledge and skills required for the safe application, usage, and removal of energy control devices.
- All employees who work in areas where energy control procedure(s) is/are utilized need to be instructed in the purpose and use of the energy control procedure(s) and in the prohibition against

attempting to restart or reenergize machines or equipment that is locked or tagged out.

- All employees who are authorized to lock out machines or equipment and perform service and maintenance operations need to be trained to recognize applicable hazardous energy sources in the workplace, the types and magnitudes of energy found in the workplace, and the means and methods of isolating and/or controlling the energy.
- All employees must know the specific procedures and limitations relating to tagout systems.
- All employees must receive regular refresher training to maintain their proficiency and to learn new or changed control methods.

Sample Lockout/Tagout Worksheet

Equipment or Process: ____
Location of Equipment: ____
A tag is required on each Isolation Location listed below.
The *specific* Type of Lockout Device must be used at the location listed.
Date prepared: ____
Prepared by: ____

Type of Energy	Isolation Location	Type of Lockout Device
Electrical		
Potential (stored)		
Kinetic (in motion)		
Pneumatic (air-gas pressure)		
Hydraulic		
Thermal		
Chemical		
Special Hazards	**Procedure for Control of Special Hazard**	

Special Procedures

Stored Energy Release Procedure

Notes
Isolation Location should specifically identify the breaker, valve, switch, or other disconnect or blocking device to be locked and tagged in order to isolate the source of energy from the work area.
Type of Lockout should specifically name the type of locking device—e.g., breaker clip, valve handwheel cover, blank flange, etc.—needed to ensure that the disconnect or blocking device stays in the isolated condition/position.
Stored Energy: Following the application of the lockout or tagout devices to the energy-isolating devices, all potential or residual energy will be reduced, disconnected, restrained, or otherwise rendered safe.

BASICS OF MACHINE SAFEGUARDING

Many hazards are created by moving, live, and energized machines. Therefore, safeguards are essential to protect workers from needless and preventable injuries.

A good principle to remember is that any machine part, function, or process that may cause injury must be safeguarded. When the operation of or accidental contact with a machine can injure the operator or others in the vicinity, the hazards must be either controlled or eliminated.

- Dangerous moving parts in three basic areas require safeguarding:
  - The point of operation: that point where work—such as cutting, shaping, boring, or forming of stock—is performed on material.
  - Power transmission apparatus: all components of the mechanical system that transmit energy to the part of the machine performing the work. These components include flywheels, pulleys, belts, connecting rods, couplings, cams, spindles, chains, cranks, and gears.
  - Other moving parts: all parts of the machine that

move while the machine is in use. These can include reciprocating, rotating, and transverse moving parts as well as feed mechanisms and auxiliary parts.

HAZARDOUS MECHANICAL MOTIONS AND ACTIONS

A wide variety of mechanical motions and actions may constitute hazards to the worker. These can include the movements of:
- rotating members
- reciprocating arms
- moving belts
- meshing gears
- cutting teeth
- any parts that crush or shear

The basic types of hazardous mechanical motions and actions are:
- Motions
  - rotating (including in-running nip points)
  - reciprocating
  - transversing
- Actions
  - cutting
  - punching
  - shearing
  - bending

Motions

Rotating motions can be dangerous; even smooth, slowly rotating shafts can grip clothing and, through mere skin contact, force an arm or hand into a dangerous position. Injuries due to contact with rotating parts can be severe.

Collars, couplings, cams, clutches, flywheels, shaft ends, spindles, meshing gears, and horizontal or vertical shafting are some examples of common rotating mechanisms that can be hazardous. The danger increases when projections such as set screws, bolts, nicks, abrasions, and keys are exposed on rotating parts.

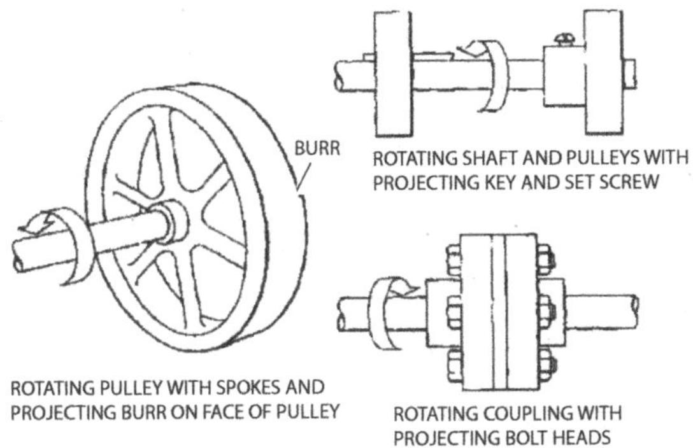

In-running nip point hazards are caused by the rotating parts of machinery. There are three main types of in-running nips:

- Parts can rotate in opposite directions while their axes are parallel to each other. These parts may be in contact (producing a nip point) or in close proximity. In the latter case, the stock fed between the rolls produces the nip points. This danger is common with machines that have intermeshing gears, rolling mills, and calenders.

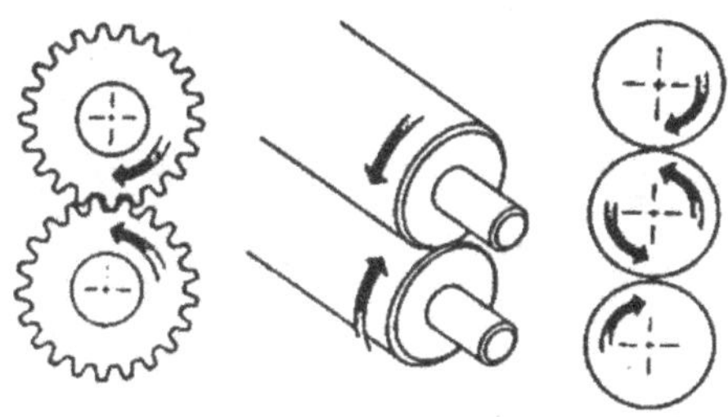

- Nip points are also created between rotating and tangentially moving parts. Some examples are the point of contact between a power transmission belt and its pulley, a chain and a sprocket, and a rack and a pinion. See the figure.

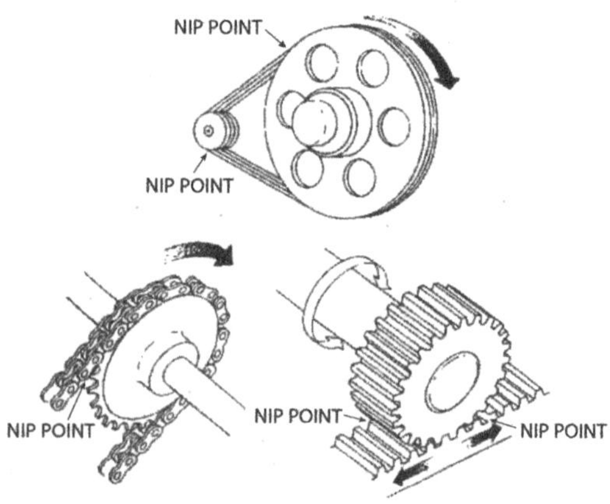

- Nip points can develop between rotating and fixed parts that create a shearing, crushing, or abrading motion. Examples are spoked handwheels or flywheels, screw conveyors, the periphery of an abrasive wheel, and an incorrectly adjusted work rest.

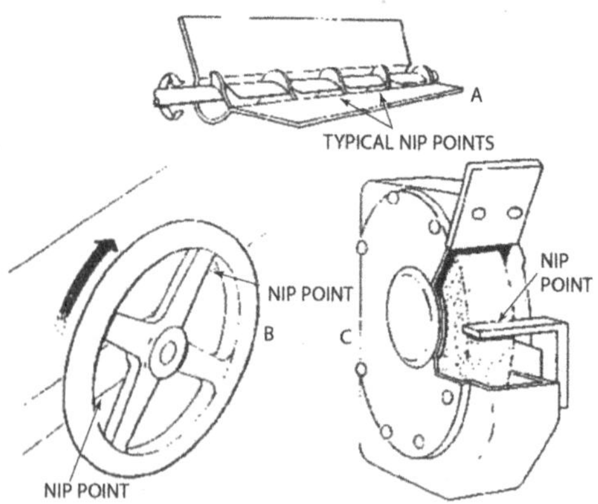

Reciprocating motions can be hazardous because, during the back-and-forth or up-and-down motions, a worker may be struck by or caught between a moving part and a stationary part.

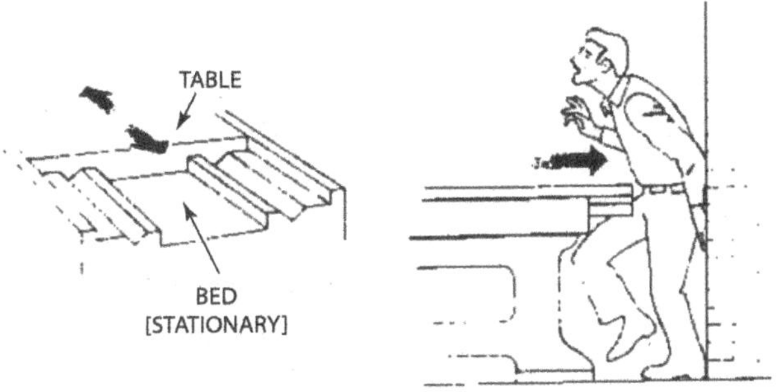

Transverse motion (movement in a straight, continuous line) creates a hazard because a worker may be struck or caught in a pinch or shear point by a moving part.

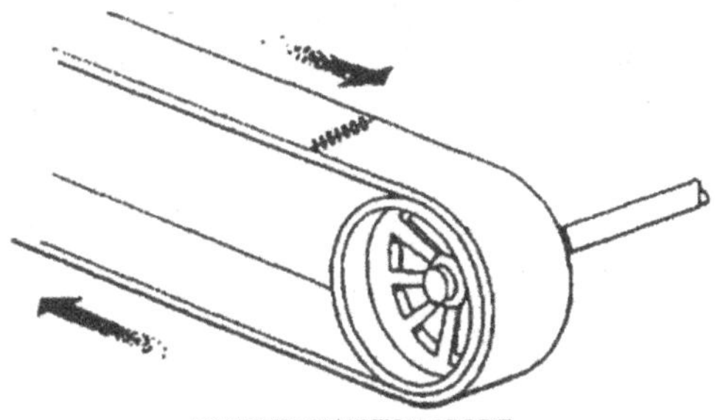

Actions

Cutting actions may involve rotating, reciprocating, or transverse motions. The danger of cutting actions exists at the point of operation when cutting wood, metal, or other materials, where finger, arm, and body injuries can occur and where flying chips or scrap material can strike the head, perhaps in the eyes or face. Examples of machinery that incorporates cutting hazards include band saws, circular saws, boring or drilling machines, turning machines (lathes), and milling machines. See the figure.

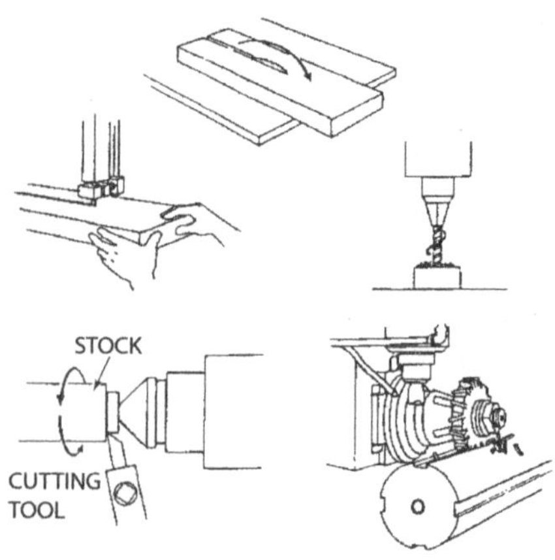

Punching actions result when power is applied to a slide (ram) for the purpose of blanking, drawing, or stamping metal or other materials. The danger from this type of action occurs at the point of operation where stock is inserted, held, and withdrawn by hand. Typical machines used for punching operations are power presses and iron works.

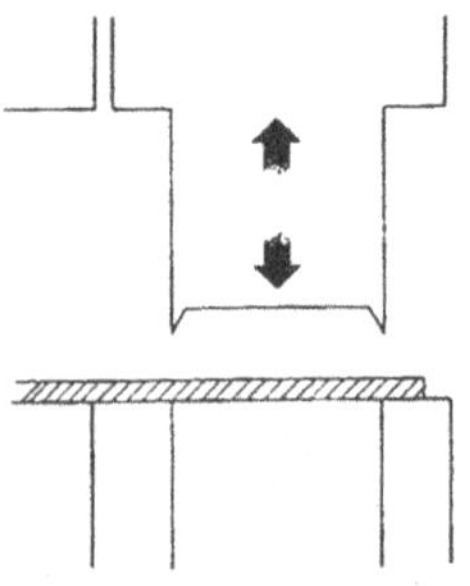

Shearing actions involve applying power to a slide or knife in order to trim or shear metal or other materials. A hazard is present at the point of operation where stock is manually inserted, held, and withdrawn. Examples of machinery used for shearing operations are mechanically, hydraulically, and pneumatically powered shears.

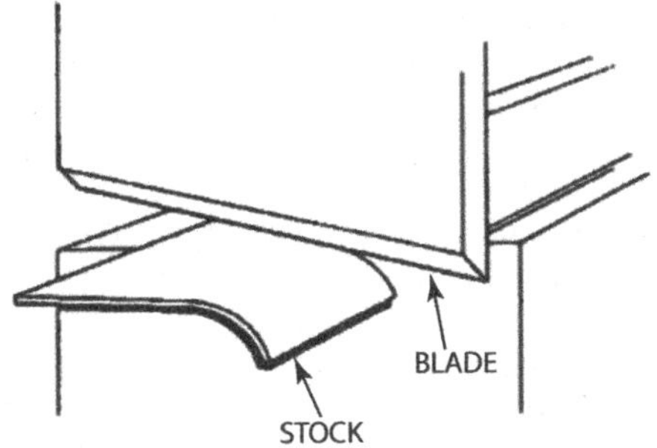

Bending actions result when power is applied to a slide in order to draw or stamp metal or other materials. A hazard can occur at the point of operation where stock is inserted, held, and withdrawn. Equipment that involves bending actions includes power presses, press brakes, and tubing benders.

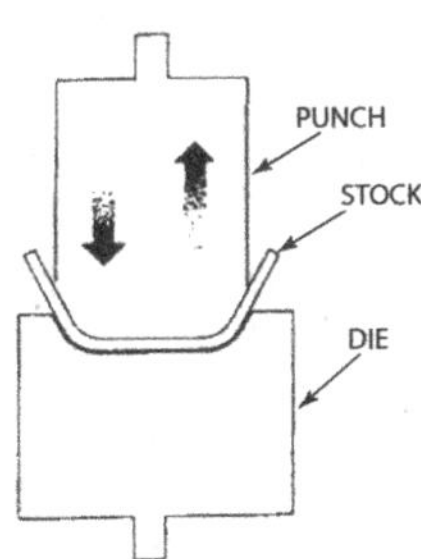

Requirements for Safeguards

Safeguards must meet the following minimum general requirements:

- Prevent contact—Safeguards must prevent hands, arms, and other parts of a worker's body from making contact with moving parts. A good safeguarding system eliminates the possibility of an operator or other worker placing parts of his or her body near moving parts.
- Be secure—Workers should not be able to easily remove or tamper with the safeguard; a safeguard

that can easily be made ineffective is no safeguard at all. Guards and safety devices should be made of durable materials that will withstand the conditions of normal use. They must also be firmly attached to the machine.

- Shield from falling objects—The safeguard should ensure that no objects can fall onto moving parts. A small tool that is dropped into a cycling machine could easily become a projectile that could strike and injure someone.
- Create no new hazards—A safeguard defeats its own purpose if it creates a hazard of its own, such as a shear point, jagged edge, or unfinished surface that can cause a laceration. The edges of guards, for instance, should be rolled or bolted in such a way that sharp edges are eliminated.
- Create no interference—Any safeguard that prevents a worker from performing a job quickly and without difficulty might soon be overridden or disregarded. Proper safeguarding can actually enhance efficiency since it can reduce a worker's apprehensions about possible injury.
- Allow for safe lubrication—If possible, a worker should be able to lubricate the machine without needing to remove the safeguards. Being able to situate oil reservoirs outside the guard, with a tube leading to the lubrication point, will reduce the need for the operator or maintenance worker to get near the hazardous area.

ELECTRICITY

Basic electrical formulas provide the information necessary to determine the amount of electricity in a line or in the human body. They are critical to understanding the nature of a catastrophic failure or accident. Ohm's law is the basic electrical formula; many other calculations are derived from it.

Bonding and Grounding

Bonding is a method of connecting two objects so that the static charge and potential difference are eliminated. Grounding provides a path from any container to earth.

Capacitance

Capacitance = calculation of energy dissipated

$E = \frac{1}{2}cv^2$

E = energy in millijoules

c = capacitance in picofarads (10^{12})

v = volts $\times 10^{-9}$

GFCI

A ground fault circuit interrupter (GFCI) detects 5 milliamps of electricity in 1/40th of a second. It detects a current change to the ground, not phase to phase. When the change is detected, an internal device stops the current flowing through it. A charge of 5 milliamps may produce a muscle spasm in a human, and only 100 milliamps is generally fatal.

Hazard Classifications

The *National Electric Code* classifies locations as to their use. Different uses have different ratings based on their potential hazards. Here are NEC Article 500 requirements for electrical equipment:

Class	Division 1	Division 2
I. Flammable gases, vapors, and liquids	Present in ignitable/explosive concentrations	Not normally in explosive concentrations
II. Dusts	Ignitable quantities present	Not normally in ignitable quantities
III. Fibers and fine particles	Material actively handled or used in manufacturing	In storage or handled in storage

Inductance

$L_{series} = L_1 + L_2 + L_3....$
$L_{parallel} = 1/L_1 + 1/L_2 + 1/L_3....$

Joule's Law

Power $= I^2R$
Power = current squared × resistance

Ohm's Law

$E = I \times R$
E = voltage
I = current
R = resistance

Ohm's Wheel

Parallel Circuits

$C_{parallel} = C_1 + C_2 + C_3...$
$\quad C$ = capacitance
$E_{parallel} = E_1 = E_2 = E_3...$
$\quad E$ = voltage
$I_{parallel} = I_1 + I_2 + I_3...$
$\quad I$ = current
$1/R_{parallel} = 1/R_1 + 1/R_2 + 1/R_3...$
$\quad R$ = resistance

Power

$P = EI$
Power (watts) = volts × current
Power (watts) = volts × volts/ohms
Power (watts) = amps × amps ohms

Resistivity Calculations

$R = \rho \times (L/A)$
R = resistivity
ρ = resistivity in Ω per unit length
L = length
A = area

$R = 1/\sigma$
• R = resistivity
• σ = conductivity

Rubber Insulating Equipment Voltage Requirements

Class of Equipment	Maximum Use Voltage[1] a-c-rms	Retest Voltage[2] a-c-rms	Retest Voltage[2] d-c-avg
0	1,000	5,000	20,000
1	7,500	10,000	40,000
2	17,000	20,000	50,000
3	26,500	30,000	60,000
4	36,000	40,000	70,000

[1]The maximum use voltage is the a–c voltage (rms) classification of the protective equipment that designates the maximum nominal design voltage of the energized system that may be safely worked. The nominal design voltage is equal to the phase-to-phase voltage on multiphase circuits. However, the phase-to-ground potential is considered to be the nominal design voltage:
(1) If there is no multiphase exposure in a system area and if the voltage exposure is limited to the phase-to-ground potential, or
(2) If the electrical equipment and devices are insulated, isolated, or both so that the multiphase exposure on a grounded wye circuit is removed.
[2]The proof-test voltage should be applied continuously for at least 1 minute, but no more than 3 minutes.

Safety Clearances

Because of electricity's effect on equipment and materials, safety guidelines have been developed to protect people and maintain equipment's integrity. The following charts outline various electrical clearance distances and testing requirements. Crane clearances of electrical lines are 10 ft for 0–50 kV + ¼ in. for each kV over 50 kV.

Approach Distance for Qualified Employees

Voltage Range (Phase to Phase)	Minimum Approach Distance
300 V and less	Avoid Contact
Over 300 V, not over 750 V	1 ft 0 in. (30.5 cm)
Over 750 V, not over 2 kV	1 ft 6 in. (46 cm)
Over 2 kV, not over 15 kV	2 ft 0 in. (61 cm)
Over 15 kV, not over 37 kV	3 ft 0 in. (91 cm)
Over 37 kV, not over 87.5 kV	3 ft 6 in. (107 cm)
Over 87.5 kV, not over 121 kV	4 ft 0 in. (122 cm)
Over 121 kV, not over 140 kV	4 ft 6 in. (137 cm)

Series Circuits

- $C_{series} = 1/C_1 + 1/C_2 + 1/C_3...$
  - C = capacitance
- $E_{series} = E_1 + E_2 + E_3...$
  - E = voltage
- $I_{series} = I_1 = I_2 = I_3...$
  - I = current
- $R_{series} = R_1 + R_2 + R_3...$
  - R = resistance

Electrical Symbols

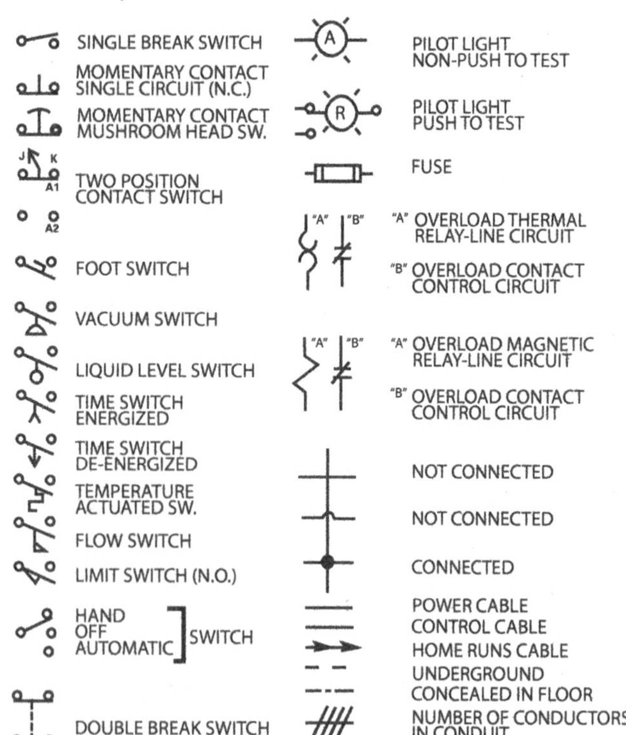

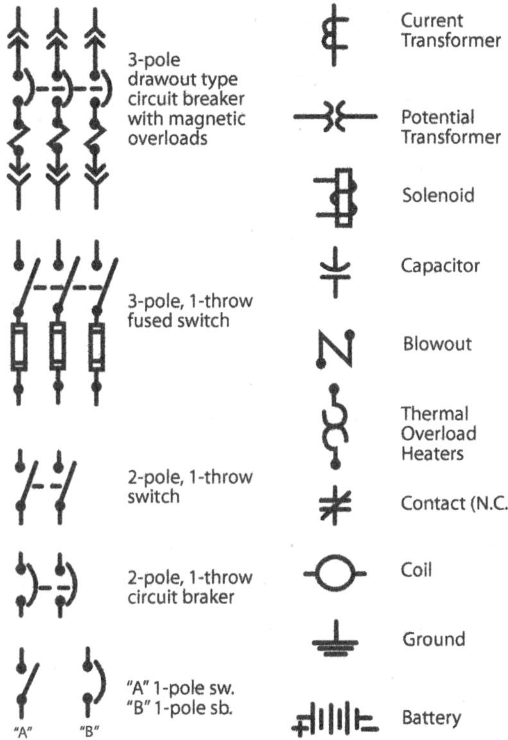

LIGHT

Light—or, more precisely, lack of light—is found to be the root cause of many accidents every year. People lose their footing, miss a step, or fall from an elevation because of low light levels. Often, industrial shops need better lighting levels because of the machinery in operation. First aid rooms need even better lighting levels to accurately examine injuries. OSHA has issued lighting level guidelines for both general industry and construction.

Minimum Illumination Intensities in Foot-Candles

Foot-Candles	Area of Operation
5	General construction area lighting
3	General construction areas, concrete placement, excavation and waste areas, access ways, active storage areas, loading platforms, refueling areas, and field maintenance areas
5	Tunnels, shafts, and general underground work areas. (Exception: A minimum of 10 foot-candles is required at tunnel and shaft headings during drilling, mucking, and scaling. Bureau of Mines–approved cap lights must be acceptable for use in the tunnel heading.)
5	Indoors: warehouses, corridors, hallways, and exit ways
10	General construction plant and shops (e.g., batch plants, screening plants, mechanical and electrical equipment rooms, carpenter shops, rigging lofts and active storerooms, mess halls, indoor toilets, and workrooms)
30	First aid stations, infirmaries, and offices

For areas or operations not covered above, refer to the American National Standard ANSI/IES RP7–2001, *Practice for Industrial Lighting*, for recommended values of illumination.

Formulas

$n = c/v$

n = index of refraction

c = speed of light in air

v = speed of light in material

$I_2 = I_1[(d_1)^2/(d_2)^2]$

intensity at location$_2$ = intensity at location$_1$ × (distance to location$_1$)^2/(distance to location$_2$)^2

Physical Properties

Flux—light traveling through a unit of area

Foot-candle (fc)—lumen/ft^2

Foot-candle (fc)—10.76 candela/m^2 = 10.76 lux

Illumination—measured in foot-candles (English units) or lux (metric units)

100 lumens at/on 1 ft^2 of area =
100 fc of illumination
100 lumens at/on 1 m^2 of area = 100 lux

Intensity—measured in candela (candlepower emitted from source)

Lumen—approximately equivalent to candela

Luminance—measured in foot-lamberts (light emitted/reflected per unit of area) or candela/m^2

Output (lumens)—light from 1 candle falling on 1 ft^2 1 foot from the candle

Physical brightness—measured in foot-lamberts

Rate of flow (luminous flux)—measured in lumens

Reflectance—measured in percentage (%)

Speed of light—186,000 miles/sec or $2.9978 \times (10)^{10}$ cm/sec

Visible Light Spectrum

Ultraviolet (UV)	100 to 380–400 nm
UV-C	100 to 280 nm
UV-B	280 to 315–320 nm
UV-A	315–320 to 380–400 nm
Visible Light	380–400 to 760–780 nm
Infrared (IR)	760–780 nm to 1 mm
IR-A	760–780 nm to 1.4 μm
IR-B	1.4–3.0 μm
IR-C	3.0 μm to 1 mm

RADIATION

Radiation can be of two distinct types: ionizing and non-ionizing. Ionizing radiation is ion-producing energy. An ion is a form of an element that has an electron charge different from the usual charge. The charge may be either positive or negative, depending upon whether the element has gained or lost electrons. An isotope is a form of an element in which the number of electrons and protons is the usual number, but the number of neutrons has changed. Isotopes emit several different types of particles or rays.

Workers are exposed most often to two types of non-ionizing radiation: ultraviolet (UV) and infrared (IR). UV radiation can cause sunburn (erythema), keratitis, and conjunctivitis. IR radiation creates heat and can cause cataracts.

- Ionizing = has the energy to strip off electrons from neutron atoms
- Nonionizing (microwaves, TV remotes) = can't strip electrons
  - Ultraviolet (UV) = wavelength, not energy
  - Infrared (IR) = light exposure (passes through cornea and into retina)
- Alpha particle = no hazard except when inhaled
- Beta particle = external and internal hazards
- Gamma particle = comes from something happening first
- Curie = activity of decay

Type	Effects	Shielding
Alpha	Short range—< 4" in air Chemically similar to calcium (can collect in kidneys, bones, liver, lungs, and spleen) Eyes represent an internal exposure.	Skin, paper, thin film of water
Beta	Secondary release of gamma radiation Higher energies can cause skin burns.	Light metals (like aluminum)
Neutron	Secondary release of gamma radiation	Carbon or high hydrogen content (like water)
Gamma and X ray	Most penetrating Electromagnetic radiation—gamma (natural) and x ray (manmade)	Heavy metals (like lead)

Radiation Protection—Time, Distance, and Shielding

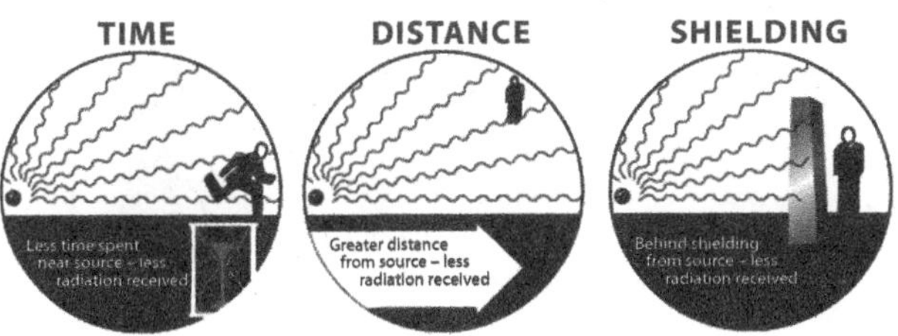

Reproduced with permission from the NDT Resource Center and The Center for NDE, Iowa State University.

- Alpha particles are the least powerful of the types of particles and consist of a helium nucleus. Alpha particles may be stopped by air, clothing, or paper. They are an inhalation hazard but are too weak to penetrate a film badge. Some alpha emitters can mimic calcium and may accumulate in the bones.
- Beta particles are more powerful than alpha particles. They may be stopped by plastic, lead shielding, or water. They can be detected by a film badge.
- Gamma rays are emitted from the nucleus. More substantive shielding, such as lead shielding, concrete, water, or iron, is needed for protection. They have a short wavelength.
- X rays are basically electron orbits changing places. A Geiger counter can detect all types of ionizing radiation but especially beta, gamma, and x rays.
- Exposure is controlled by limiting the emissions at the source, limiting the time a worker is exposed, increasing the distance to the source, or by shielding.
- The amount of radiation at a distance follows the inverse square law, where the intensity of exposure is inversely related to the distance.

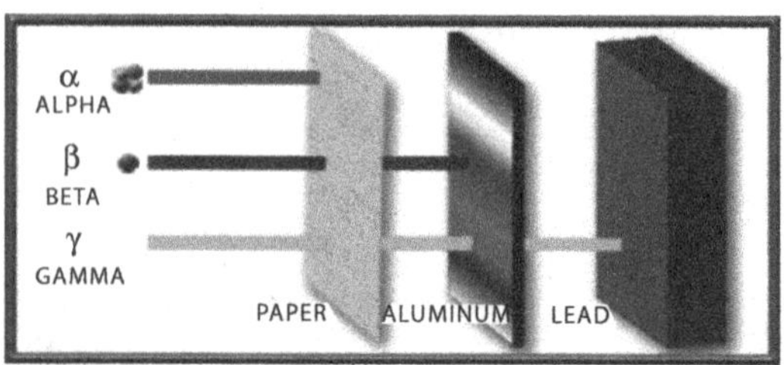

Inverse Square Law

$I_1/I_2 = D_1{}^2/D_2{}^2$

  I_1 = intensity at location$_1$
  I_2 = intensity at location$_2$
  D_1 = (distance to location$_1$)^2
  D_2 = (distance to location$_2$)^2

- A curie is the activity of radiated material decay. It is based on radium and was calculated by Marie Curie to be 3.7×10^{10} disintegrations/sec.
- A rad is a dose of radiation: the energy absorption per mass of tissue.
  - 1 rad = absorption of 100 ergs/gram, 10^{-2} grays
- A roentgen is a measure of ionized air, which is calculated to be
  - 2.58×10^{-4} coulomb/kg of air
- A rem is the measure of a dose of radiation and its biological effect on living tissue relative to 1 roentgen of x rays. It is calculated to be 10^{-2} sieverts. A worker is allowed to be exposed to 5 rems per 12 months, up to five occasions. The Department of Transportation limits exposure of workers to 200 millirem/hour from a package surface, while a cab driver may be exposed only to 2 millirem/hour.
  - A rem has several equivalents:
    - = dose of 1 roentgen from x rays or gamma rays
    - = 1 rad of x, gamma, or beta rays
    - = 0.1 rad from neutrons or high-energy particles
    - = 0.05 rad from particles heavier than protons and that have enough energy to reach the eye lens

FORMULAS

Source
$S \cong 6CE$
radiation reading = 6 roentgen hours at 1 foot × curies × radiation energy in MeV (megaelectron volts)

Calculation for Half-Lives or Other-Lives
$T = (-\ln \frac{1}{2})/K$
Half-life = $(-\ln$ of $\frac{1}{2})$/the disintegration constant for that element
Third-life = $(-\ln$ of $\frac{1}{3})$/K, and so forth, for other-life calculations

Calculation for Nuclear Decay
$N_t = N_0 e^{-kt}$
Number of atoms at the end of the decay period = number of atoms at the beginning of the decay period
$e^{-\text{disintegration constant} \times \text{time}}$

- Nonionizing radiation is energy that does not produce or release ions. From a worker's safety point of view, lasers, microwaves, and sunlight are the most common exposures.
- Laser (light amplification by stimulated emission of radiation) is a light beam of high energy classified by its power level. Class 1 lasers are low power; Class 4 lasers are high power. The colors of warning signs are magenta (purple) on yellow.

Employees who use hand or power tools and equipment encounter many hazards—falling, flying, abrasive, and splashing objects and harmful dusts, fumes, mists, vapors, and gases—and must be provided with the appropriate safeguards necessary to protect themselves from the hazards. All hazards involved in the use of power tools can be mitigated by following five basic safety rules:

- Use regular maintenance to keep all tools in good condition.
- Use the right tool for the job.
- Examine each tool for damage before use.
- Operate the tool according to the manufacturer's instructions.
- Use the proper protective equipment.

COMPRESSED GASES

The most common use of compressed gases in the workplace is for welding, although many laboratories use nitrogen in test equipment.

Acetylene gas is 92.3% carbon and 7.7% hydrogen (H_2). Its flammable range is an impressive 2.5%–81%. Oxygen tanks are typically pressurized to 2,200 psi, and hydrogen tanks are similarly pressurized to 2,000 psi.

Oxygen cylinders should be kept 20 ft away from flammable gas cylinders or located on the other side of a 30-minute fire wall.

Gas regulators and hoses have a standardized color scheme: Red is the color for the fuel gas; green is the color for oxygen.

The regulators and hoses need to be approved by Underwriters Laboratories (UL) or Factory Mutual. LPG hoses have a minimum working pressure of 250 psig and a burst pressure of 1,250 psig.

CONFINED SPACES

Confined spaces must be carefully monitored, as they may be dangerous locations because of insufficient oxygen, engulfment, toxic fumes, or poor ventilation. A workplace should be inspected for the presence of confined spaces, and procedures should be established, including those for restricted entry, hazards identification, written rescue plans, and written contingency plans.

Individuals who enter a confined space are called, naturally, entrants. Those who monitor the entry from outside the confined space are called attendants. Each must receive training appropriate to his or her duties.

A permit system should be developed to assess the hazards of each entry. The permit is a written document for the specific entry and lists the entrants and attendants, hazards, isolation means, rescuers, and air quality. Each permit is unique to each entry and is valid for only one project. If the project extends over a shift or over days, a new permit should be created at the beginning of each new workshift. Sometimes it is possible to reclassify a permit-required confined space as a non-permit-required confined space if the hazards have been eliminated.

CRANES

Cranes are used to move very heavy loads. Good crane operators have a feel for the equipment that is advantageous. They must be able to judge both horizontal and vertical distances, air movement, and load angles in order to deliver the load to the target. No crane operator works alone, however, and the operator's skill is affected by the quality of the spotter.

Many of the crane's mechanical limits are on the crane's date plate. The crane load radius is the distance from the center of rotation to the center of the load after boom deflection. Typically, cranes are able to lift 125% of their maximum intended load, but cranes should lift this much only in the case of emergencies. All other loads must weigh less than the manufacturer's stated maximum intended load. If two or more cranes share the same runway, they should be kept 30 ft apart. All cranes or overhead equipment must kept at least 10 ft from electrical lines to avoid accidental contact and potential electrocution.

EXCAVATIONS

An excavation is a trench greater than 15 ft wide, whereas other types of trenches have a depth greater than their width, with 15 ft as their maximum width. Trench and excavation cave-ins generally have catastrophic results because a cubic yard of soil may weigh as much as 2,000 lb. When a company is preparing to begin a project, many states and municipalities require that notification to underground facility organizations be given 24 hours before the start of earthmoving activities. Some states require 48-hour notification. This notification allows the underground facility organizations time to alert those utilities with underground facilities in the area to mark the locations of their facilities.

All spoils (i.e., the earth removed from the trench) must be kept back at least 2 ft from the edge of the trench. Inspections must be performed by a qualified person before anyone enters the trench, during work activities, after work is completed, after rainstorms, or whenever conditions change. A ladder or other means of egress is required if the trench is 4 ft deep or greater, with a maximum horizontal distance to the ladder of 25 ft.

To protect workers from cave-ins, trenches are either sloped or shored. Shoring (or shielding) is the use of wood or metal structures to hold back the edges of the trench. Sloping is the beveling back of the edges of the trench from the bottom of the trench to the grass level. A special case is a trench less than 5 ft deep in stable rock, where neither sloping nor shoring is necessary.

The required angle of the slope depends on the type of soil encountered. A hard, cohesive soil can withstand a steeper angle than a soft, sandy soil can. Determination of soil type is explained in 29 *CFR* 1926, Subpart P.

Type A soil is cohesive and can withstand higher forces. Type C soil is the most unstable soil type. Type B soil is in the middle. If the soil type cannot be determined, then the requirements of Type C soil must be followed because they are the most restrictive and require the removal of the greatest amount of soil from the trench. The allowable slopes are outlined below.

Table B–1. Maximum Allowable Slopes

Soil or Rock Type	Maximum Allowable Slopes (H:V)[1] for Excavations Less than 20 ft deep[3]
Stable Rock	Vertical (90°)
Type A[2]	¾:1 (53°)
Type B	1:1 (45°)
Type C	1½:1 (34°)

Footnote (1) Numbers shown in parentheses next to maximum allowable slopes are angles expressed in degrees from the horizontal. Angles have been rounded off.

Footnote (2) A short-term maximum allowable slope of ½H:1V (63 degrees) is allowable in excavations in Type A soil that is 12 ft (3.67 m) or less in depth. Short-term maximum allowable slopes for excavations greater than 12 ft (3.67 m) in depth shall be ¾H:1V (53 degrees).

Footnote (3) Sloping or benching for excavations greater than 20 ft deep shall be designed by a registered professional engineer.

Source: 29 *CFR* 1926, Appendix B, Subpart P.

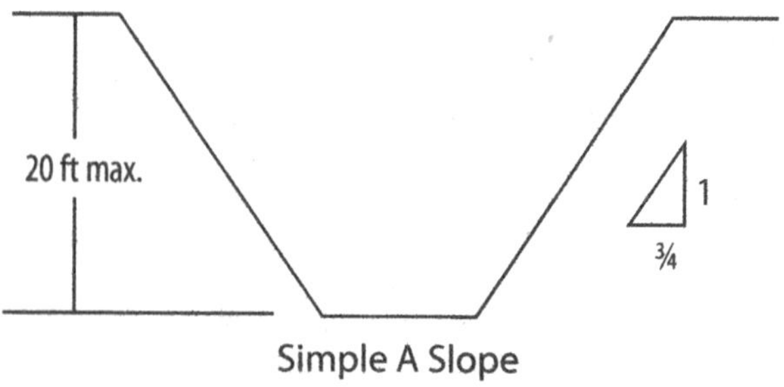

Simple A Slope

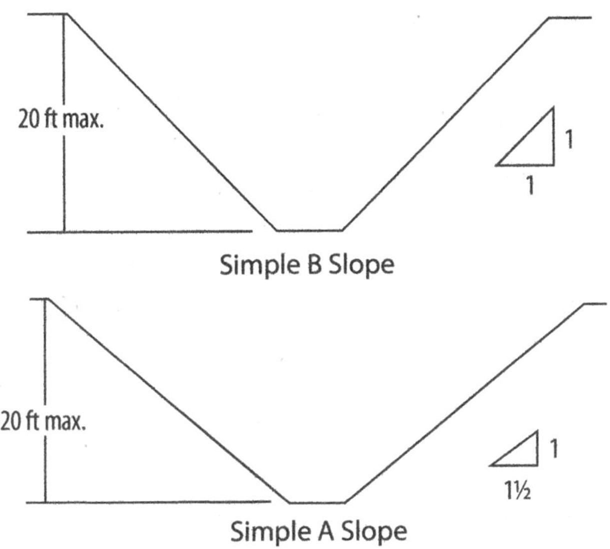

Rather than excavate massive amounts of soil, many contractors use shoring or shielding systems. Aluminum systems are light, easy to work with, and compact. The following charts provide the required component needs and parameters. Shoring (or shores) are vertical structures that hold back earth. Walers are horizontal structures that hold back earth. The charts may be confusing in distance requirements—remember that for shores, the distances indicated are the horizontal spacing of the vertical component, whereas for walers, the distance is the vertical distance requirement for the horizontal component.

Although their construction is more time-consuming than that of aluminum systems, wooden shoring systems are still used in many excavations. Wooden shoring systems are complicated by the type of timber used for the components. Charts are included for each type of soil (A, B, and C) and whether the timber components are fir or oak. The fir charts indicate S4S (surfaced four sides) for sizing of the timber; those for oak indicate actual timber size.

Aluminum Hydraulic Shoring

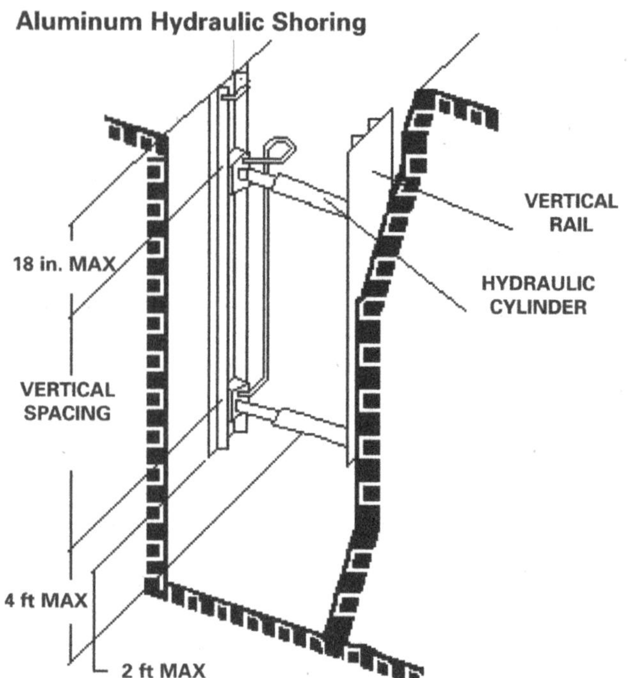

Vertical Aluminum Hydraulic Shoring
(Spot Bracing)

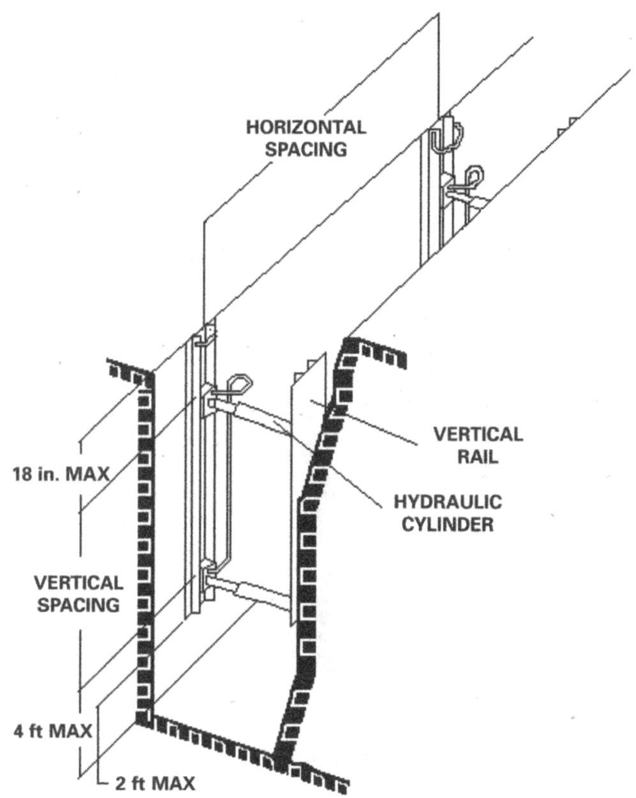

Vertical Aluminum Hydraulic Shoring
(With Plywood)

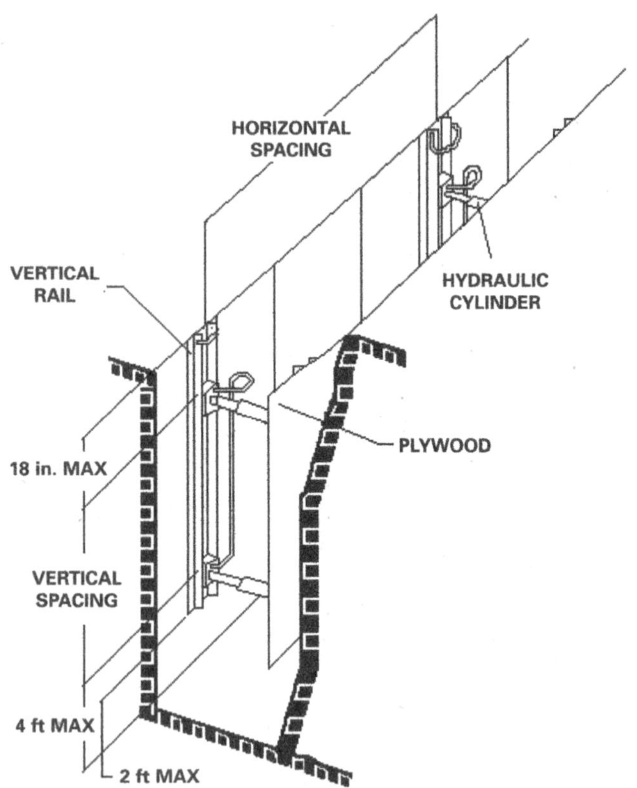

Vertical Aluminum Hydraulic Shoring (Stacked)

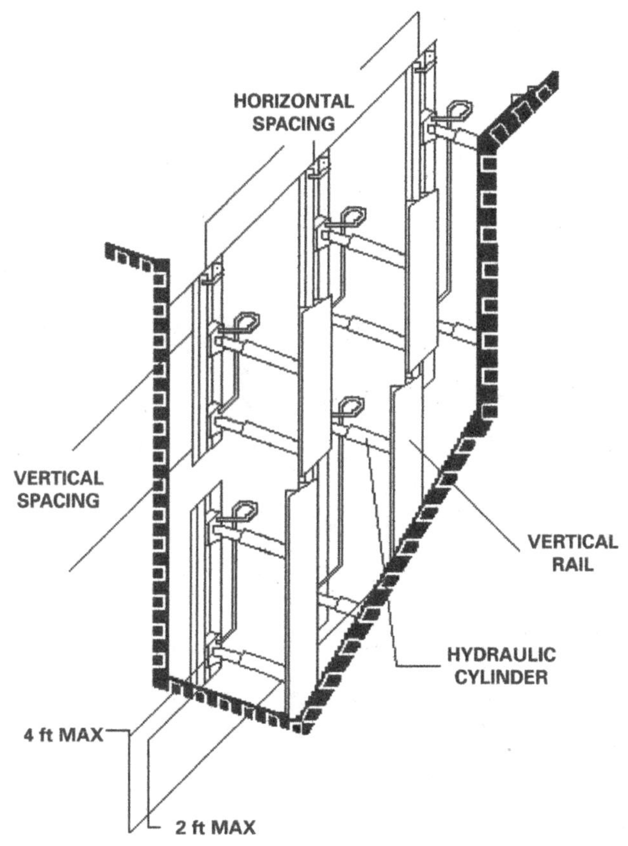

Aluminum Hydraulic Shoring Waler System (Typical)

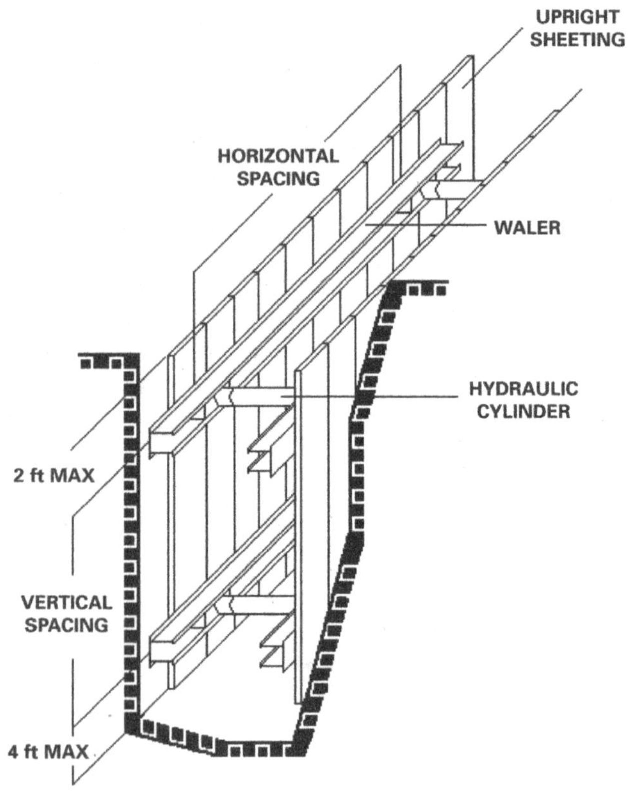

Pneumatic Shoring

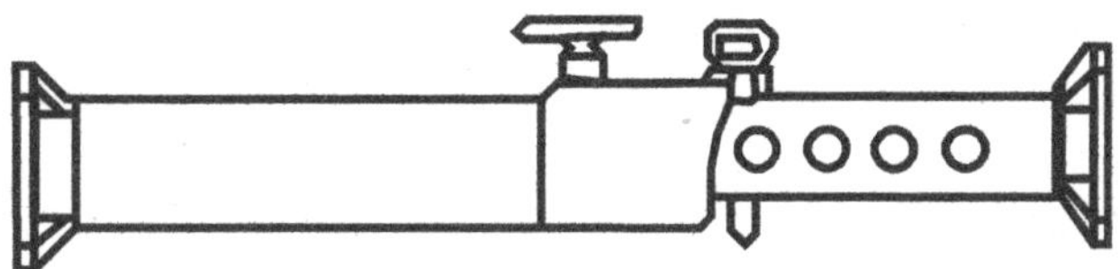

Jacks and Shields

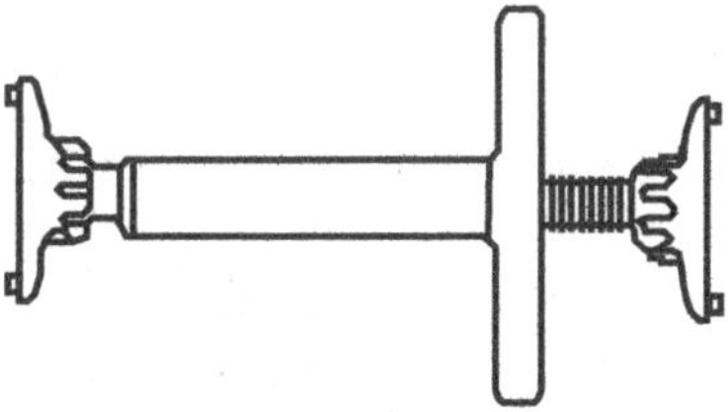

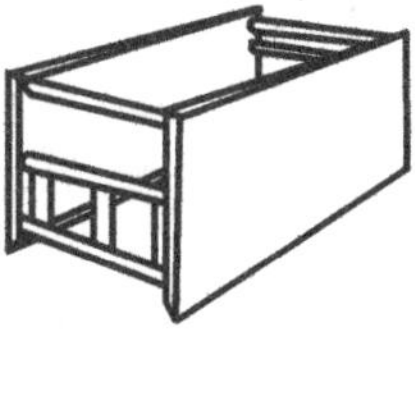

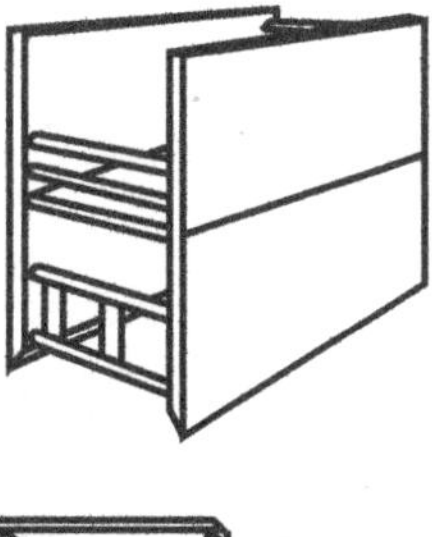

Aluminum Hydraulic Shoring Vertical Shores for Soil Type A

DEPTH OF TRENCH (FEET)	MAXIMUM HORIZONTAL SPACING (FEET)	MAXIMUM VERTICAL SPACING (FEET)	HYDRAULIC CYLINDERS		
			WIDTH OF TRENCH (FEET)		
			UP TO 8	OVER 8 UP TO 12	OVER 12 UP TO 15
OVER 5 UP TO 10	8	4	2 INCH DIAMETER	2 INCH DIAMETER NOTE (2)	3 INCH DIAMETER
OVER 10 UP TO 15	8				
OVER 15 UP TO 20	7				
OVER 20	NOTE (1)				

Footnotes to tables, and general notes on hydraulic shoring, are found in *CFR* 1926, Subpart P, Appendix D, Item (g)
Note (1): See *CFR* 1926, Subpart P, Appendix D, Item (g) (1)
Note (2): See *CFR* 1926, Subpart P, Appendix D, Item (g) (2)

Aluminum Hydraulic Shoring Vertical Shores for Soil Type B

DEPTH OF TRENCH (FEET)	MAXIMUM HORIZONTAL SPACING (FEET)	MAXIMUM VERTICAL SPACING (FEET)	HYDRAULIC CYLINDERS		
			WIDTH OF TRENCH (FEET)		
			UP TO 8	OVER 8 UP TO 12	OVER 12 UP TO 15
OVER 5 UP TO 10	8	4	2 INCH DIAMETER	2 INCH DIAMETER NOTE (2)	3 INCH DIAMETER
OVER 10 UP TO 15	6.5				
OVER 15 UP TO 20	5.5				
OVER 20	NOTE (1)				

Footnotes to tables, and general notes on hydraulic shoring, are found in *CFR* 1926, Subpart P, Appendix D, Item (g)
Note (1): See *CFR* 1926, Subpart P, Appendix D, Item (g) (1)
Note (2): See *CFR* 1926, Subpart P, Appendix D, Item (g) (2)

Aluminum Hydraulic Shoring — Water Systems for Soil Type B

DEPTH OF TRENCH	WALES		HYDRAULIC CYLINDERS						TIMBER UPRIGHTS		
	VERTICAL SPACING	SECTION MODULES *	WIDTH OF TRENCH (FEET)						MAX. HORIZ. SPACING (ON CENTER)		
			UP TO 8		OVER 8 UP TO 12		OVER 12 UP TO 15				
(FEET)	(FEET)	(INCHES)	HORIZ. SPACING	CYLINDER DIAMETER	HORIZ. SPACING	CYLINDER DIAMETER	HORIZ. SPACING	CYLINDER DIAMETER	SOLID SHEET	2 FEET	3 FEET
OVER 5 UP TO 10	4	3.5	8.0	2 INCHES	8.0	2 INCHES NOTE (2)	8.0	3 INCHES	—	—	3 × 12
		7.0	9.0	2 INCHES	9.0	2 INCHES NOTE (2)	9.0	3 INCHES			
		14.0	12.0	3 INCHES	12.0	3 INCHES	12.0	3 INCHES			
OVER 10 UP TO 15	4	3.5	6.0	2 INCHES	6.0	2 INCHES NOTE (2)	6.0	3 INCHES	—	3 × 12	—
		7.0	8.0	3 INCHES	8.0	3 INCHES	8.0	3 INCHES			
		14.0	10.0	3 INCHES	10.0	3 INCHES	10.0	3 INCHES			
OVER 15 UP TO 20	4	3.5	5.5	2 INCHES	5.5	2 INCHES NOTE (2)	5.5	3 INCHES	3 × 12	—	—
		7.0	6.0	3 INCHES	6.0	3 INCHES	6.0	3 INCHES			
		14.0	9.0	3 INCHES	9.0	3 INCHES	9.0	3 INCHES			
OVER 20	NOTE (1)										

Footnotes to tables, and general notes on hydraulic shoring, are found in *CFR* 1926, Subpart P, Appendix D, Item (g)

Note (1): See *CFR* 1926, Subpart P, Appendix D, Item (g) (1)

Note (2): See *CFR* 1926, Subpart P, Appendix D, Item (g) (2)

* Consult product manufacturer and/or qualified engineer for Section Modulus of available wales.

Aluminum Hydraulic Shoring — Water Systems for Soil Type C

DEPTH OF TRENCH	WALES		HYDRAULIC CYLINDERS						TIMBER UPRIGHTS		
	VERTI-CAL SPACING	SECTION MOD-ULES *	WIDTH OF TRENCH (FEET)						MAX. HORIZ. SPACING (ON CENTER)		
			UP TO 8		OVER 8 UP TO 12		OVER 12 UP TO 15				
(FEET)	(FEET)	(INCHES)	HORIZ. SPACING	CYLINDER DIAMETER	HORIZ. SPACING	CYLINDER DIAMETER	HORIZ. SPACING	CYLINDER DIAMETER	SOLID SHEET	2 FEET	3 FEET
OVER 5 UP TO 10	4	3.5	6.0	2 INCHES	6.0	2 INCHES NOTE (2)	6.0	3 INCHES	3 × 12	—	—
		7.0	6.5	2 INCHES	6.5	2 INCHES NOTE (2)	6.5	3 INCHES			
		14.0	10.0	3 INCHES	10.0	3 INCHES	10.0	3 INCHES			
OVER 10 UP TO 15	4	3.5	4.0	2 INCHES	4.0	2 INCHES NOTE (2)	4.0	3 INCHES	3 × 12	—	—
		7.0	5.5	3 INCHES	5.5	3 INCHES	5.5	3 INCHES			
		14.0	8.0	3 INCHES	8.0	3 INCHES	8.0	3 INCHES			
OVER 15 UP TO 20	4	3.5	3.5	2 INCHES	3.5	2 INCHES NOTE (2)	3.5	3 INCHES	3 × 12	—	—
		7.0	5.0	3 INCHES	5.0	3 INCHES	5.0	3 INCHES			
		14.0	6.0	3 INCHES	6.0	3 INCHES	6.0	3 INCHES			
OVER 20	NOTE (1)										

Footnotes to tables, and general notes on hydraulic shoring, are found in *CFR* 1926, Subpart P, Appendix D, Item (g)

Note (1): See *CFR* 1926, Subpart P, Appendix D, Item (g) (1)

Note (2): See *CFR* 1926, Subpart P, Appendix D, Item (g) (2)

* Consult product manufacturer and/or qualified engineer for Section Modulus of available wales.

Timber Trench Shoring — Minimum Timber Requirements* Soil Type A $P_a = 25 \times h \pm 72$ psf (2 ft Surcharge)

DEPTH OF TRENCH (FEET)	SIZE (S4S) AND SPACING OF MEMBERS**													
	CROSS BRACES							WALES		UPRIGHTS				
	HORIZ. SPACING (FEET)	WIDTH OF TRENCH (FEET)					VERT. SPACING (FEET)	SIZE (IN.)	VERT. SPACING (FEET)	MAXIMUM ALLOWABLE HORIZONTAL SPACING (FEET)				
		UP TO 4	UP TO 6	UP TO 9	UP TO 12	UP TO 15				CLOSE	4	5	6	8
5	UP TO 6	4×4	4×4	4×4	4×4	4×6	4	NOT REQ'D	NOT REQ'D				4×6	
TO	UP TO 8	4×4	4×4	4×4	4×6	4×6	4	NOT REQ'D	NOT REQ'D					4×8
	UP TO 10	4×6	4×6	4×6	6×6	6×6	4	8×8	4			4×8		
10	UP TO 12	4×6	4×6	4×6	6×6	6×6	4	8×8	4				4×6	
10	UP TO 6	4×4	4×4	4×4	6×6	6×6	4	NOT REQ'D	NOT REQ'D				4×10	
TO	UP TO 8	4×6	4×6	4×6	6×6	6×6	4	6×8	4		4×6			
	UP TO 10	6×6	6×6	6×6	6×6	6×6	4	8×8	4			4×8		
15	UP TO 12	6×6	6×6	6×6	6×6	6×6	4	8×10	4		4×6		4×10	
15	UP TO 6	6×6	6×6	6×6	6×6	6×6	4	6×8	4	3×6				
TO	UP TO 8	6×6	6×6	6×6	6×6	6×6	4	8×8	4	3×6	4×12			
	UP TO 10	6×6	6×6	6×6	6×6	6×8	4	8×10	4	3×6				
20	UP TO 12	6×6	6×6	6×6	6×8	6×8	4	8×12	4	3×6	4×12			
OVER 20	SEE 29 *CFR* 1926, Subpart P, Appendix C, paragraph g, Item 1.													

* Douglas fir or equivalent with a bending strength not less than 1,500 psi.
** Manufactured members of equivalent strength may be substituted for wood.

Timber Trench Shoring — Minimum Timber Requirements* Soil Type A $P_a = 25 \times h + 72$ psf (2 ft Surcharge)

DEPTH OF TRENCH (FEET)	HORIZ. SPACING (FEET)	CROSS BRACES — WIDTH OF TRENCH (FEET) UP TO 4	UP TO 6	UP TO 9	UP TO 12	UP TO 15	VERT. SPACING (FEET)	WALES SIZE (IN.)	WALES VERT. SPACING (FEET)	UPRIGHTS — MAX. ALLOWABLE HORIZONTAL SPACING (FEET) CLOSE	4	5	6	8
5	UP TO 6	4 × 4	4 × 4	4 × 6	6 × 6	6 × 6	4	NOT REQ'D	—				2 × 6	
TO	UP TO 8	4 × 4	4 × 4	4 × 6	6 × 6	6 × 6	4	NOT REQ'D	—					2 × 8
	UP TO 10	4 × 6	4 × 6	4 × 6	6 × 6	6 × 6	4	8 × 8	4			2 × 6		
10	UP TO 12	4 × 6	4 × 6	6 × 6	6 × 6	6 × 6	4	8 × 8	4				2 × 6	
10	UP TO 6	4 × 4	4 × 4	4 × 6	6 × 6	6 × 6	4	NOT REQ'D	—				3 × 8	
TO	UP TO 8	4 × 6	4 × 6	6 × 6	6 × 6	6 × 6	4	6 × 8	4		2 × 6			
	UP TO 10	6 × 6	6 × 5	6 × 6	6 × 8	6 × 8	4	8 × 8	4			2 × 6		
15	UP TO 12	6 × 6	6 × 6	6 × 6	6 × 8	6 × 8	4	8 × 10	4				3 × 8	
15	UP TO 6	6 × 6	6 × 6	6 × 6	6 × 8	6 × 8	4	6 × 8	4	3 × 6				
TO	UP TO 8	6 × 6	6 × 6	6 × 6	6 × 8	6 × 8	4	8 × 8	4	3 × 6				
	UP TO 10	8 × 8	8 × 8	8 × 8	8 × 8	8 × 10	4	8 × 10	4	3 × 6				
20	UP TO 12	8 × 8	6 × 6	8 × 8	8 × 8	8 × 10	4	8 × 12	4	3 × 6				
OVER 20	SEE 29 *CFR* 1926, Subpart P, Appendix C, paragraph g, Item 1.													

* Mixed oak or equivalent with a bending strength not less than 850 psi.

** Manufactured members of equivalent strength may be substituted for wood.

Timber Trench Shoring — Minimum Timber Requirements* Soil Type B $P_a = 45 \times h + 72$ psf (2 ft Surcharge)

DEPTH OF TRENCH (FEET)	SIZE (ACTUAL) AND SPACING OF MEMBERS**													
	CROSS BRACES							WALES		UPRIGHTS				
	HORIZ. SPACING (FEET)	WIDTH OF TRENCH (FEET)					VERT. SPACING (FEET)	SIZE (IN.)	VERT. SPACING (FEET)	MAXIMUM ALLOWABLE HORIZONTAL SPACING (FEET)				
		UP TO 4	UP TO 6	UP TO 9	UP TO 12	UP TO 15				CLOSE	2	3	4	6
5 TO 10	UP TO 6	4×6	4×6	4×6	6×6	6×6	5	6×8	5			3×12 4×8		4×12
	UP TO 8	4×6	4×6	6×6	6×6	6×6	5	8×8	5		3×8		4×8	
	UP TO 10 SEE NOTE 1	4×6	4×6	6×6	6×6	6×8	5	8×10	5			4×8		
10 TO 15	UP TO 6	6×6	6×6	6×6	6×8	6×8	5	8×8	5	3×6	4×10			
	UP TO 8	6×8	6×8	6×8	8×8	8×8	5	10×10	5	3×6	4×10			
	UP TO 10 SEE NOTE 1	6×8	6×8	8×8	8×8	8×8	5	10×12	5	3×6	4×10			
15 TO 20	UP TO 6	6×8	6×8	6×8	6×8	8×8	5	8×10	5	4×6				
	UP TO 8	6×8	6×8	6×8	8×8	8×8	5	10×12	5	4×6				
	UP TO 10 SEE NOTE 1	8×8	8×8	8×8	8×8	8×8	5	12×12	5	4×6				
OVER 20	SEE 29 *CFR* 1926, Subpart P, Appendix C, paragraph g, Item 1.													

* Douglas fir or equivalent with a bending strength not less than 1,500 psi.

** Manufactured members of equivalent strength may be substituted for wood.

Timber Trench Shoring — Minimum Timber Requirements* Soil Type B $P_a = 45 \times h + 72$ psf (2 ft Surcharge)

DEPTH OF TRENCH (FEET)	HORIZ. SPACING (FEET)	CROSS BRACES						WALES		UPRIGHTS		
		WIDTH OF TRENCH (FEET)					VERT. SPACING (FEET)	SIZE (IN.)	VERT. SPACING (FEET)	MAXIMUM ALLOWABLE HORIZONTAL SPACING (FEET)		
		UP TO 4	UP TO 6	UP TO 9	UP TO 12	UP TO 15				CLOSE	2	3
5 TO 10	UP TO 6	4×6	4×6	6×6	6×6	6×6	5	6×8	5			2×6
	UP TO 8	6×6	6×6	6×6	6×8	6×8	5	8×10	5			2×6
	UP TO 10 SEE NOTE 1	6×6	6×6	6×6	6×8	6×8	5	10×10	5			2×6
10 TO 15	UP TO 6	6×6	6×6	6×6	6×8	6×8	5	8×8	5		2×6	
	UP TO 8	6×8	6×8	6×8	8×8	8×8	5	10×10	5		2×6	
	UP TO 10 SEE NOTE 1	8×8	8×8	8×8	8×8	8×10	5	10×12	5		2×6	
15 TO 20	UP TO 6	6×8	6×8	6×8	8×8	8×8	5	8×10	5	3×6		
	UP TO 8	8×8	8×8	8×8	8×8	8×10	5	10×12	5	3×6		
	UP TO 10 SEE NOTE 1	8×10	8×10	8×10	8×10	10×10	5	12×12	5	3×6		
OVER 20	SEE 29 *CFR* 1926, Subpart P, Appendix C, paragraph g, Item 1.											

* Mixed oak or equivalent with a bending strength not less than 850 psi.
** Manufactured members of equivalent strength may be substituted for wood.

Timber Trench Shoring — Minimum Timber Requirements* Soil Type C $P_a = 80 \times h + 72$ psf (2 ft Surcharge)

DEPTH OF TRENCH (FEET)		SIZE (ACTUAL) AND SPACING OF MEMBERS**												
		CROSS BRACES							WALES		UPRIGHTS			
	HORIZ. SPACING (FEET)	WIDTH OF TRENCH (FEET)					VERT. SPACING (FEET)	SIZE (IN.)	VERT. SPACING (FEET)	MAXIMUM ALLOWABLE HORIZONTAL SPACING (FEET)				
		UP TO 4	UP TO 6	UP TO 9	UP TO 12	UP TO 15				CLOSE				
5	UP TO 6	6×6	6×6	6×6	6×6	8×6	5	8×8	5	3×6				
TO	UP TO 8	6×6	6×6	6×6	8×8	8×8	5	10×10	5	3×6				
	UP TO 10 SEE NOTE 1	6×6	6×6	8×8	8×8	8×8	5	10×12	5	3×6				
10														
10	UP TO 6	6×8	6×8	6×8	8×8	8×8	5	10×10	5	4×6				
TO	UP TO 8 SEE NOTE 1 SEE NOTE 1	8×8	8×8	8×8	8×8	8×8	5	12×12	5	4×6				
15														
15	UP TO 6 SEE NOTE 1 SEE NOTE 1 SEE NOTE 1	8×8	8×8	8×8	8×10	8×10	5	10×12	5	4×6				
TO														
20														
OVER 20	SEE 29 *CFR* 1926, Subpart P, Appendix C, paragraph g, Item 1.													

*   Douglas fir or equivalent with a bending strength not less than 1,500 psi.
** Manufactured members of equivalent strength may be substituted for wood.

Timber Trench Shoring — Minimum Timber Requirements* Soil Type C $P_a = 80 \times h + 72$ psf (2 ft Surcharge)

DEPTH OF TRENCH (FEET)	SIZE (ACTUAL) AND SPACING OF MEMBERS**														
	CROSS BRACES							WALES		UPRIGHTS					
	HORIZ. SPACING (FEET)	WIDTH OF TRENCH (FEET)					VERT. SPACING (FEET)	SIZE (IN.)	VERT. SPACING (FEET)	MAXIMUM ALLOWABLE HORIZONTAL SPACING (FEET)					
		UP TO 4	UP TO 6	UP TO 9	UP TO 12	UP TO 15				CLOSE					
5	UP TO 6	6 × 8	6 × 8	6 × 8	8 × 8	8 × 8	5	8 × 10	5	2 × 6					
TO	UP TO 8	8 × 8	8 × 8	8 × 8	8 × 8	8 × 10	5	10 × 12	5	2 × 6					
10	UP TO 10 SEE NOTE 1	8 × 10	8 × 10	8 × 10	8 × 10	10 × 10	5	12 × 12	5	2 × 6					
10	UP TO 6	8 × 8	8 × 8	8 × 8	8 × 8	8 × 10	5	10 × 12	5	2 × 6					
TO 15	UP TO 8 SEE NOTE 1 SEE NOTE 1	8 × 10	8 × 10	8 × 10	8 × 10	10 × 10	5	12 × 12	5	2 × 6					
15 TO 20	UP TO 6 SEE NOTE 1 SEE NOTE 1 SEE NOTE 1	8 × 10	8 × 10	8 × 10	8 × 10	10 × 10	5	12 × 12	5	3 × 6					
OVER 20	SEE 29 *CFR* 1926, Subpart P, Appendix C, paragraph g, Item 1.														

* Mixed oak or equivalent with a bending strength not less than 850 psi.
** Manufactured members of equivalent strength may be substituted for wood.

GUARDING AND SAFETY DEVICES

Guarding provides the final physical barrier between a worker and a machine's danger zone. Types of protection are summarized below.

Types of Protection

Distance	Keep out of reach
Guards	
Enclosure/fixed guards	Barrier, keeps hands out and material in
Interlock guard	Open the guard and the power shuts off
Adjustable	Change the opening
Self -adjusting	Guards change with material; shape, then change back to original shape after material passes
Location	Put hazardous machinery where there are no people
Safety devices	
Presence sensing	Photoelectric, radio frequency
Pullback	Cables attached to operator, retract
Restraint	Similar to pullback but with fixed length
Safety trip	Pressure-sensitive bar
Two-hand control	Both hands on during entire cycle
Two-hand trip	Two hands start cycle

The Safety Distance Formula is a calculation that determines whether sufficient distance is built into the guarding arrangement of a machine that has sensing devices to prevent an operator's hands from entering the danger zone during operation. For more information on the formula, see 29 *CFR* 1910.217.

$D = 63$ in./sec $\times\ t$

D = inches from point of operation, the minimum safety distance in inches

t = stopping time of machine in seconds

63 in./sec = assumed hand speed constant

GRINDERS

For bench grinders, the work rest is a maximum of ⅛ in. from the wheel, and the tongue guard is ¼ in. away. The wheel guard should be adjustable and cover as much of the wheel as possible. Ring test the wheel before installing it by suspending it from a pencil (or other such axis) and lightly tapping it. A clear-sounding ring indicates no cracks, and a leaden sound indicates a defect.

Surface speed of rotating equipment (ft per min)—e.g., the wheel on a grinder:

$$SFM = \pi \times D \times rpm$$

$$SFM = \pi \times \text{diameter of the wheel (ft)} \times \text{revolutions per minute}$$

HAZARD COLORS

Standardized color schemes allow the worker to discern information about his or her surroundings at a glance. For instance, orange indicates machine guards; magenta indicates radiation hazards. These colors are used on signs, labels, barrier tape, and other informational media.

Color Code for Hazard Type

Red	Fire, danger, emergency stop
Red orange	Vehicular traffic hazards
Green	Safety, first aid equipment
Magenta	Radiation hazards
Yellow	Tripping hazards, caution
Black/white	Housekeeping, traffic
Orange	Dangerous parts of machinery
Orange red	Biohazard
Blue	Information signs

Reference: After 29 *CFR* 1910. 144. ANSIZ53.1.

Labels have a hierarchy of the degree of severity of a hazard. They are standardized and reflect the color scheme indicated above.

Label Meanings

Caution label	Possibility of minor personal injury, product damage, or property damage, usually yellow
Warning label	Possibility of death or severe personal injurt, usually orange
Danger label	Immediate hazard, death or severe personal injury probable, usually red

LUMBER

Lumber has its own terms and definitions. Lumber prices vary greatly depending on the quality and the amount of preparation or dressing the mill has performed. As stated previously, the excavation standard has different requirements for fir (S4S) and oak (actual or rough cut).

Calculation for Board Feet of Lumber

1×4	Divide length in feet by 3
1×6	Divide length in feet by 2
1×8	Multiply length in feet by ⅔
1×12	Length in feet = board feet
2×4	Multiply length in feet by ⅔
2×6	Length in feet = board feet
2×8	Multiply length in feet by 1⅓
2×12	Multiply length in feet by 2

Lumber Dressing

Rough Lumber	Surfaced show marks
Surfaced Lumber (dressed)	
S1S	Surfaced 1 side
S1E	Surfaced 1 edge
S2S	Surfaced 2 sides
S2E	Surfaced 2 edges
S1S1E	Surfaced 1 side, 1 edge
S1S2E	Surfaced 1 side, 2 edges
S2S1E	Surfaced 2 sides, 1 edge
S4S	Surfaced 4 sides
Shop Lumber/Factory Lumber	Millwork lumber
Yard Lumber	Structural lumber

NAILS

Nails come in a myriad of shapes and sizes. The commonly used nails and the number of nails per pound of a standard wire nail are shown in the following chart.

Common Nails

Size	Length (in.)	#/lb
2d	1	900
3d	1.25	615
4d	1.50	325
6d	2	200
8d	2.5	105
10d	3	75
12d	3.25	60
16d spike	3.5	45
20d	4	30
30d	4.5	20
40d	5	14
50d	5.5	10
60d	6	8

PIPE THREADS

Pipe threads have been standardized in the United States (American Standard Pipe Threads) and can be determined by the code on the pipe. The code indicates the nominal size, the number of threads per inch, and the thread series symbols.

Pipe Threads Codes

Symbol	Use
C	Coupling
L	Locknut
M	Mechanical
N	American Standard (National)
P	Pipe
R	Rail fittings
S	Straight
T	Taper

POWDER-ACTUATED TOOLS

Powder-actuated tools use a small charge to deliver a fastener to the substrate. The power of the charge is indicated by the combination of the case and load colors.

Powder-Actuated Tool Charges

Power Level	Case Color	Load Color
1	Brass	Gray
2	Brass	Brown
3	Brass	Green
4	Brass	Yellow
5	Brass	Red
6	Brass	Purple
7	Nickel	Gray
8	Nickel	Brown
9	Nickel	Green
10	Nickel	Yellow
11	Nickel	Red
12	Nickel	Purple

PRESSURE VESSELS

Pressure vessels are under the mandate of the American Society of Mechanical Engineers (ASME). Because the vessels are under pressure, they have specific safeguards and requirements. A rupture of a pressure vessel can cause serious harm because the compressed gas boils out of the rupture in liquid form. This is a boiling liquid, expanding vapor explosion (BLEVE).

A vessel's normal operating pressure (NOP) is its everyday pressure and load. The maximum operating pressure (MOP) is its highest pressure and load. The design pressure is the maximum operating pressure and a safety factor. The maximum allowable work pressure (MAWP) is the maximum gauge pressure at the top of the vessel when at temperature. ASME requires hydrostatic tests at 150% of MAWP.

Vessel Pressure Formula:
$P = P_0 + (d \times g/g_c \times h)$
P = pressure at the base of the vessel
P_0 = pressure at the top of the vessel
d = density of the substance in the vessel; if water, 62.4 lb/ft^3
g/g_c = the atmospheric constant
h = height of vessel

Correct all units to psi (lb/in.^{2}). If the vessel is a closed tank, then the atmospheric pressure does not affect the reading. If, however, the vessel is an open tank, then add 1 atm (14.7 psi) unless the vessel is at another pressure.

Important: The psia (absolute) is the gauge reading + 14.7, and psig (gauge) is the actual reading on the pressure gauge.

A flammable/combustible storage tank is considered a pressure vessel when the pressure is over 15 psi, which is equal to approximately 2 atms of pressure: psia = 14.7 + 15 = 30 psia = 2 atm.

ROPES

A block and tackle system is used to move or hoist a heavy object by means of equipment with a mechanical advantage. A block is basically a system of movable pulleys, with wheels known as sheaves. The tackle is the rope used as the hoisting apparatus. Generally, the mechanical advantage is the number of strands of rope supporting the movable load. (The rope that has the input force applied to it is not included in the calculation of mechanical advantage.) Adding pulleys or blocks increases the mechanical advantage but also increases friction forces. Because of friction, no more than four block and tackle systems or sets of pulleys are usually used. A fixed pulley has no mechanical advantage but merely changes the direction of the input force.

Line Load

Although the comparative weight to lift is the total weight of the load divided by the number of strands or rope parts, a more specific line load calculation will also account for friction.

$$\text{line load} = \text{load}/\text{sheaves} \times (1 + \text{friction})^{\text{sheaves}}$$

Ropes are rated for their safety factor. If the safety factor is not known, it can generally be assumed to be 5. To be more conservative, a safety factor of 10 may be used. The safety factor is the margin divided by the maximum intended load that the rope may carry. It is NOT to be used as the normal intended load rating. As a calculation, the safety factor of a rope is the tensile strength divided by the safe working load. If the safety factor is 10, then the safe working load is one-tenth of the tensile strength. Nylon rope has 2.5 times the breaking strength of manila rope and 4 times the elasticity.

Breaking strength calculation (synthetic and manila rope, not wire rope):

$$B = [W + (RWS/100)/S] \times F$$

B = rope breaking strength

W = weight lifted
R = percent friction per rope part
S = number of parts of sheaves
F = safety factor

$B = W/S \times$ safety factor, if friction is ignored

Wire Rope

Wire rope is its own special case. Specific parameters exist for connections and removal from service. Wire rope is connected to itself by the use of U bolts. The curved part of the U bolt is in contact with the dead (cut) end of the wire, and the saddle is in contact with the live end.

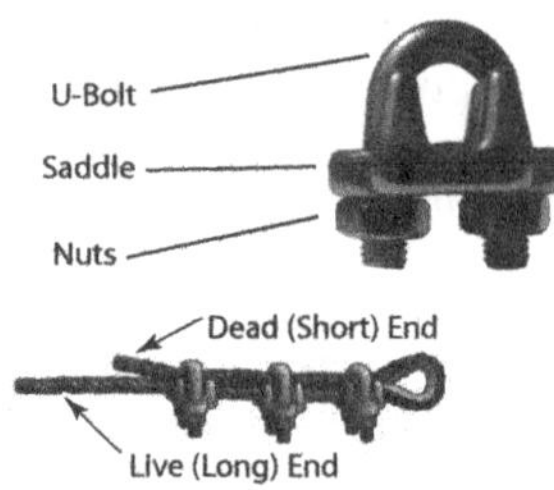

Used with permission from E-Rigging.com.

The saying, "Don't ride a dead horse" is a useful memory jogger. The saddle does not ride on the dead end. The curved part of the U bolt will dig into the wire to help hold it in place. This weakens the wire, but since the wire is in contact with the dead end, this weakening is permissible.

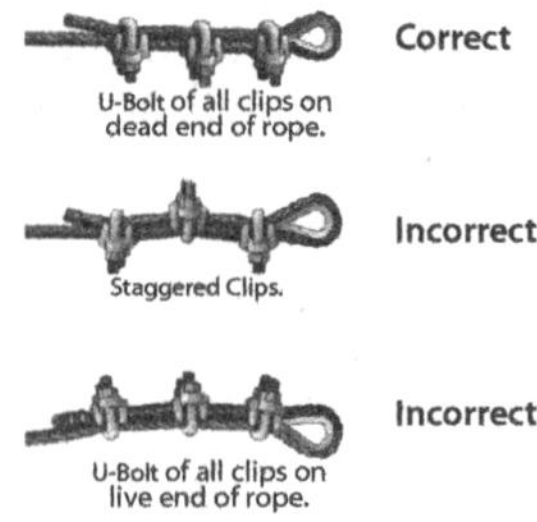

Used with permission from E-Rigging.com.

Number and Spacing of U-Bolt Wire Rope Clips

Improved plow steel, rope diameter (inches)	Number of Clips		Minimum Spacing (inches)
	Drop Forged	Other Materials	
½	3	4	3
⅝	3	4	3¾
¾	4	5	4½
⅞	4	5	5¼
1	5	6	6
1⅛	6	6	6¾
1¼	6	7	7½
1⅜	7	7	8¼
1½	7	8	9

As just noted, the wires in wire rope will deteriorate and break over time. When performing an inspection, if in a 10-diameter-equivalent length more than 10% of the wires are broken, the rope must be taken out of service. Or if 5 wires in a strand or 10 random wires in a lay are broken, the rope must be discarded.

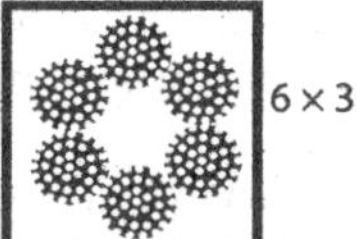

Safe Working Loads for Shackles
(In tons of 2,000 pounds)

Material size (inches)	Pin diameter (inches)	Safe working load
½	⅝	1.4
⅝	¾	2.2
¾	⅞	3.2
⅞	1	4.3
1	1⅛	5.6
1⅛	1¼	6.7
1¼	1⅜	8.2
1⅜	1½	10.0
1½	1⅝	11.9
1¾	2	16.2
2	2¼	21.2

SCAFFOLDS

Scaffolds are a deceptively intricate elevating system. Small scaffolds are very common on construction sites and within manufacturing plants. Large scaffolding systems are an art in themselves, and the designer of such a system needs training, education, and skill.

Scaffolding requirements are covered in both the General Industry (29 *CFR* 1910) and the Construction standards (29 *CFR* 1926), with some contradictions. Some basic pointers about smaller scaffolds include the following:

- Guardrails are required at 42 in. from the working platform.
- Guardrails must be able to withstand 200 lb of force on horizontal members with minimum deflection.
- Guardrails are required when the working platform is 6 ft from the ground.
- Guardrails are required on scaffolds 4 to 10 ft high with a 45-in. minimum base dimension.
- Scaffolds must be designed to withstand four times the maximum intended load, although if they are supplemented with rope or wire supports, they may support six times the intended load.
- Use ½-in. screening to protect passersby who may travel under the scaffold.
- Use scaffold-grade lumber, not regular lumber.
- A light-duty scaffold may use boards 10 ft long and support 25 lb/ft.
- A medium-duty scaffold may use boards 8 ft long and support 50 lb/ft.
- A heavy-duty scaffold may use boards 6 ft long and support 75 lb/ft.
- Planks must overlap by 12 in.
- Planks must extend over the edge by 6 in., with a 12-in. maximum overhang.
- Toeboards must be 4 in. high.
- At heights over 25 ft, the scaffold must be secured at 25-ft intervals if it is a wooden scaffold.

- For suspended scaffolds, there must be at least three turns of wire on the drum.

Minimum Nominal Size and Maximum Spacing of Members of Single-Pole Scaffolds–Light Duty

	Maximum Height of Scaffold	
	20 feet	**60 feet**
Uniformly distributed load	Not to exceed 25 lb per square ft (PSI)	
Poles or uprights	2 by 4 in.	4 by 4 in.
Pole spacing (longitudinal)	6 ft 0 in.	10 ft 0 in.
Maximum width of scaffold	5 ft 0 in.	5 ft 0 in.
Bearers or putlogs to 3 ft 0 in. width	2 by 4 in.	2 by 4 in.
Bearers or putlogs to 5 ft 0 in. width	2 by 6 in. or 3 by 4 in.	2 by 6 in. or 3 by 4 in. (rough)
Ledgers	1 by 4 in.	1¼ by 9 in.
Planking	1¼ by 9 in. (rough)	2 by 9 in.
Vertical spacing of horizontal members	7 ft 0 in.	7 ft 0 in.
Bracing, horizontal and diagonal	1 by 4 in.	1 by 4 in.
Tie-ins	1 by 4 in.	1 by 4 in.
Toeboards	4 in. high (minimum)	4 in. high (minimum)
Guardrail	2 by 4 in.	2 by 4 in.

All members except planking area are used on edge.

Minimum Nominal Size and Maximum Spacing of Members of Single-Pole Scaffolds–Medium Duty

Uniformly distributed load	Not to exceed 50 lb per square ft
Maximum height of scaffold	60 ft.
Poles or uprights	4 by 4 in.
Pole spacing (longitudinal)	8 ft 0 in.
Maximum width of scaffold	5 ft 0 in.
Bearers or putlogs	2 by 9 in. or 3 by 4 in.
Spacing of bearers or putlogs	8 ft 0 in.
Ledgers	2 by 9 in.
Vertical spacing of horizontal members	9 ft 0 in.
Bracing, horizontal	1 by 6 in. or 1¼ by 4 in.
Bracing, diagonal	1 by 4 in.
Tie-ins	1 by 4 in.
Planking	2 by 9 in.
Toeboards	4 in. high (minimum)
Guardrail	2 by 4 in.

All members except planking area are used on edge.

Minimum Nominal Size and Maximum Spacing of Members of Single-Pole Scaffolds–Heavy Duty

Uniformly distributed load	Not to exceed 75 lb per square ft
Maximum height of scaffold	60 ft.
Poles or uprights	4 by 4 in.
Pole spacing (longitudinal)	6 ft 0 in.
Maximum width of scaffold	5 ft 0 in.
Bearers or putlogs	2 by 9 in. or 3 by 5 in. (rough)
Spacing of bearers or putlogs	6 ft 0 in.
Ledgers	2 by 9 in.
Vertical spacing of horizontal members	6 ft 6 in.
Bracing, horizontal and diagonal	2 by 4 in.
Tie-ins	1 by 4 in.
Planking	2 by 9 in.
Toeboards	4 in. high (minimum)
Guardrail	2 by 4 in.

All members except planking area are used on edge.

Minimum Nominal Size and Maximum Spacing of Members of Independent-Pole Scaffolds–Light Duty

	Maximum Height of Scaffold	
	20 feet	60 feet
Uniformly distributed load	Not to exceed 25 lb per square ft	
Poles or uprights	2 by 4 in.	4 by 4 in.
Pole spacing (longitudinal)	6 ft 0 in.	10 ft 0 in.
Pole spacing (transverse)	6 ft 0 in.	10 ft 0 in.
Maximum width of scaffold	5 ft 0 in.	5 ft 0 in.
Bearers to 3 ft 0 in. span	2 by 4 in.	2 by 4 in.
Bearers to 5 ft 0 in. span	2 by 6 in. or 3 by 4 in.	2 by 9 in. (rough) or 3 by 8 in.
Ledgers	1¼ by 4 in.	1¼ by 9 in.
Planking	1¼ by 9 in.	2 by 9 in.
Vertical spacing of horizontal members	7 ft 0 in.	7 ft 0 in.
Bracing, horizontal and diagonal	1 by 4 in.	1 by 4 in.
Tie-ins	1 by 4 in.	1 by 4 in.
Toeboards	4 in. high	4 in. high (minimum)
Guardrail	2 by 4 in.	2 by 4 in.

All members except planking area are used on edge.

Minimum Nominal Size and Maximum Spacing of Members of Independent-Pole Scaffolds– Medium Duty

Uniformly distributed load	Not to exceed 50 lb per square ft
Maximum height of scaffold	60 ft.
Poles or uprights	4 by 4 in.
Pole spacing (longitudinal)	8 ft 0 in.
Pole spacing (transverse)	8 ft 0 in.
Maximum width of scaffold	5 ft 0 in.
Bearers or putlogs	2 by 9 in. or 3 by 5 in. (rough)
Spacing of bearers	8 ft 0 in.
Ledgers	2 by 9 in.
Vertical spacing of horizontal members	6 ft 0 in.
Bracing, horizontal	1 by 6 in. or $1\frac{1}{4}$ by 4 in.
Bracing, diagonal	1 by 4 in.
Tie-ins	1 by 4 in.
Planking	2 by 9 in.
Toeboards	4 in. high (minimum)
Guardrail	2 by 4 in.

All members except planking area are used on edge.

Minimum Nominal Size and Maximum Spacing of Members of Independent-Pole Scaffolds– Heavy Duty

Uniformly distributed load	Not to exceed 75 lb per square ft
Maximum height of scaffold	60 ft.
Poles or uprights	4 by 4 in.
Pole spacing (longitudinal)	6 ft 0 in.
Pole spacing (transverse)	8 ft 0 in.
Ledgers	2 by 9 in.
Vertical spacing of horizontal members	4 ft 6 in.
Bearers	2 by 9 in. (rough)
Bracing, horizontal and diagonal	2 by 4 in.
Tie-ins	1 by 4 in.
Planking	2 by 9 in.
Toeboards	4 in. high (minimum)
Guardrail	2 by 4 in.

All members except planking area are used on edge.

Tube and Coupler Scaffolds–Light Duty

Uniformly distributed load		Not to exceed 25 psf
Post spacing (longitudinal)		10 ft 0 in.
Post spacing (transverse)		6 ft 0 in.

Working Levels	Additional Planked Levels	Maximum Height
1	8	125 ft
2	4	125 ft
3	0	91 ft 0 in.

Tube and Coupler Scaffolds–Medium Duty

Uniformly distributed load		Not to exceed 50 psf
Post spacing (longitudinal)		8 ft 0 in.
Post spacing (transverse)		6 ft 0 in.

Working Levels	Additional Planked Levels	Maximum Height
1	6	125 ft
2	0	78 ft 0 in.

Tube and Coupler Scaffolds–Heavy Duty

Uniformly distributed load		Not to exceed 75 psf
Post spacing (longitudinal)		6 ft 6 in.
Post spacing (transverse)		6 ft 0 in.

Working Levels	Additional Planked Levels	Maximum Height
1	6	125 ft

Minimum Nominal Size and Maximum Spacing of Members of Outrigger Scaffolds

	Light duty	Medium duty
Maximum scaffold load	5 ft 0 in.	50 psf
Outrigger size	2 by 4 in.	3 × 10 in.
Maximum outrigger spacing	2 by 6 in. or 3 by 4 in.	6 ft 10 in.
Planking	1 by 4 in.	2 × 9 in.
Guardrail	1¼ by 9 in. (rough)	2 × 4 in.
Guardrail uprights	7 ft 0 in.	2 × 4 in.
Toeboards (minimum)	1 by 4 in.	4 in.

Distance between Scaffolding and Uninsulated Lines

Insulated Lines Voltage	Minimum Distance	Alternatives
Less than 300 volts 300 volts to 50 kV More than 50 kV	3 ft (0.9 m) 10 ft (3.1 m) 10 ft (3.1 m) plus 0.4 in. (1.0 cm) for each 1 kV over 50 kV	2 times the length of the line insulator, but never less than 10 ft (3.1 m)
Uninsulated Lines Voltage	**Minimum Distance**	**Alternatives**
Less than 50 kV More than 50 kV	10 ft (3.1 m) 10 ft (3.1 m) plus 0.4 in. (1.0 cm) for each 1 kV over 50 kV	2 times the length of the line insulator, but never less than 10 ft (3.1 m)

SLINGS

Slings are a convenient way to stabilize a load and to allow safe lifting. If the sling legs make a small angle to the load, then the force on the sling legs is increased. To determine the tension on each leg, use the sine of the angle of the sling to the load or the cosine of half of the angle at the hook and solve using the Pythagorean theorem. To find the horizontal force on each leg, use the tangent of the angle of the sling to the load. If the jaw of the hook increases by 15% or more, it is time to remove the hook from service.

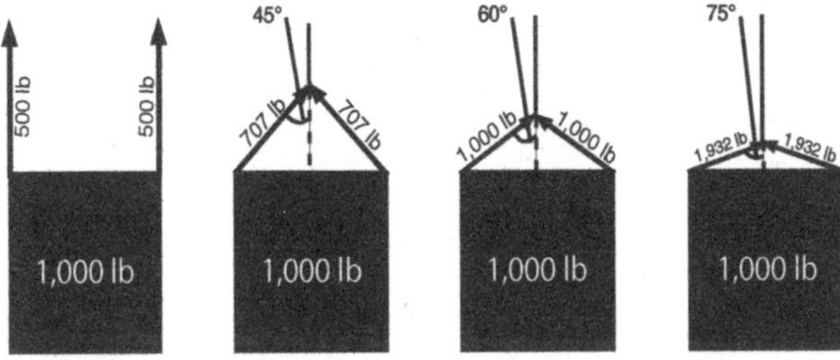

Increasing the sling angle (the angle of the sling leg with the vertical) increases the stress on each leg of the sling, even though the load remains constant. The stress on each leg can be calculated by dividing the load weight by the cosine of the angle.

Basic Sling Configurations with Vertical Legs

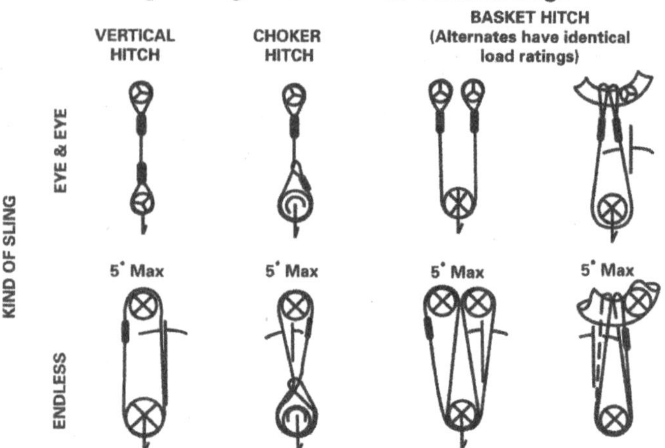

NOTES: Angles 5° or less from the vertical may be considered vertical angles.

For slings with legs more than 5° off vertical, the actual angle as shown in figure N-184-5 must be considered.

EXPLANATION OF SYMBOLS: MINIMUM DIAMETER OF CURVATURE

Represents a contact surface which shall have a diameter of curvature at least double the diameter of the rope from which the sling is made.

Represents a contact surface which shall have a diameter of curvature at least 8 times the diameter of the rope.

Represents a load in a choker hitch and illustrates the rotary force on the load and/or the slippage of the rope in contact with the load. Diameter of curvature of load surface shall be at least double the diameter of the rope.

Reference: 29 *CFR* 1910.184.

Basic Synthetic Web Sling Constructions

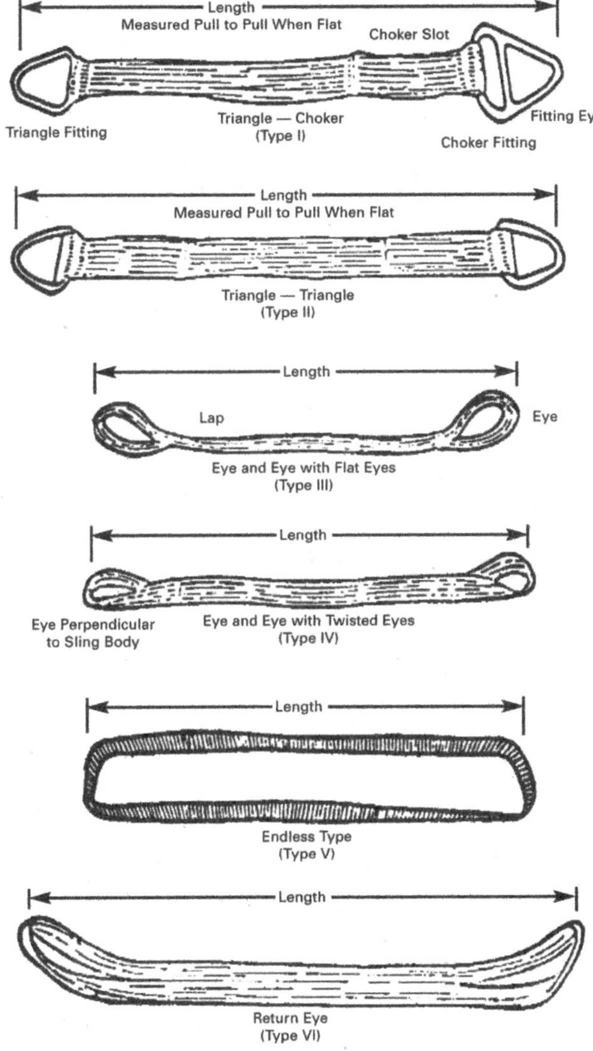

Reference: 29 *CFR* 1910.184.

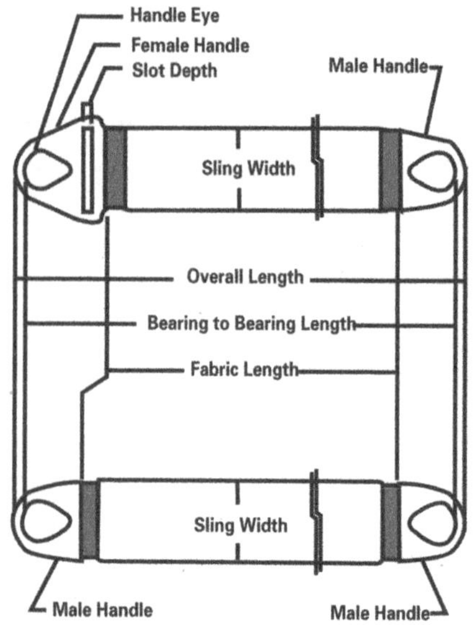

WORKING AT ELEVATION

Either a ladder or stairs must be provided if there is a change in elevation of 19 in. or more. Ladders allow workers to ascend to a height without needing stairs. This is an advantage if space is an issue (stairs take up more space) or if the height must be reached only periodically.

Ladders have several manufacturing and use requirements. The rails must be at least 11.5 in. apart, with rungs at a standard spacing (10–14 in.). When using portable ladders, the top of the ladder must extend 3 ft beyond the horizontal plane and have a 4:1 lean ratio.

Fixed ladders have their own special requirements. The maximum angle is 90°, although 75–90° is acceptable. The top of the ladder must extend 42 in. beyond the top horizontal plane and be caged beginning at 7 ft from the floor.

If stairs are provided, the steps may have only a ¼-in. variation in height. Landings are required every 12 ft (vertical), with minimum dimensions of 30 × 22 in. The landing must have a 42-in. railing. A stair rail is necessary if the stairs have four or more risers. Stair rails and handrails should be 30–34 in. high. Balusters are required every 19 in. along the stair rail. Handrails should be on the wall side of a stairway with 3 inches clearance from the wall. They are generally on the right side, going down. The handrail and the stair rail may be the same structure if they are on the open side of a stairway. A center rail is necessary if the stairs are over 88 in. wide.

Fixed industrial stairs have a 30–35° angle, although a range of 30–50° is typical for other types of stairs.

Angle of Stairway Rise

Angle to horizontal	Rise (in.)	Tread run (in.)
30° 35'	6½	11
32° 08'	6¾	10¾
33° 41'	7	10½
35° 16'	7¼	10¼
36° 52'	7½	10
38° 29'	7¾	9¾
40° 08'	8	9½
41° 44'	8¼	9¼
43° 22'	8½	9
45° 00'	8¾	8¾
46° 38'	9	8½
48° 16'	9¼	8¼
49° 54'	9½	8

Reference: Table D-1, 29 *CFR* 1910.24

A good rule of thumb for an appropriate rise and run combination is to add the height of the tread rise to the length of the tread run. The total should equal 17.5 in. Another rule of thumb is to add the run to the product of twice the rise. This total should equal 24.

To determine the number of stairs from one level to another, first find the total rise in inches, divide that by 7 to find the approximate number of tread rises, and then round up or down to a whole number to determine the number of steps. Divide the total rise by the number of steps to find the actual tread rise.

To determine the total run, multiply the individual tread run by 1 less than the total number of rises.

Where road terminology uses *grade* to indicate slope and roofs use *pitch*, stairway terminology uses *steepness*. This is based on the tangent of the angle of elevation and is the riser height divided by the tread run.

Sometimes it is impossible to provide stairs or for an individual to use stairs. In these cases, a ramp provides the means to change elevation. Two classes of ramps are mentioned in the General Industry Standards. A slope of 20° is allowable, but a slope of 15° may be preferable.

Types of Ramps

	Class A Ramp	Class B Ramp
Width	44 in. and greater	30 to 44 in.
Slope	1 to 1³⁄₁₆ in. in 12 in.	1 to 1³⁄₁₆ in. in 12 in.
Maximum height	No limit	12 ft between landings

Reference: OSHA 1910.37

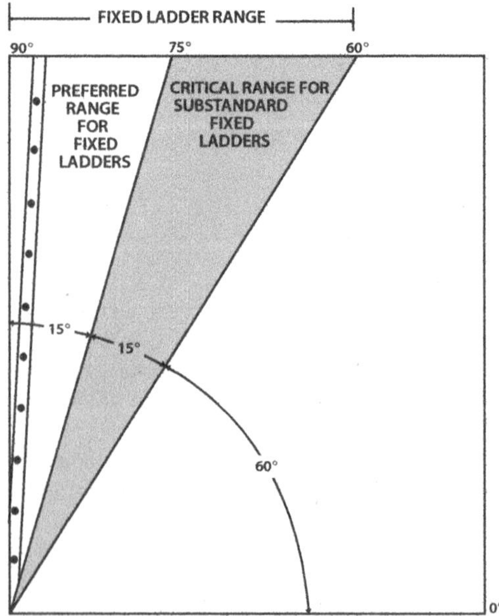

Cages for ladders more than 20 feet high

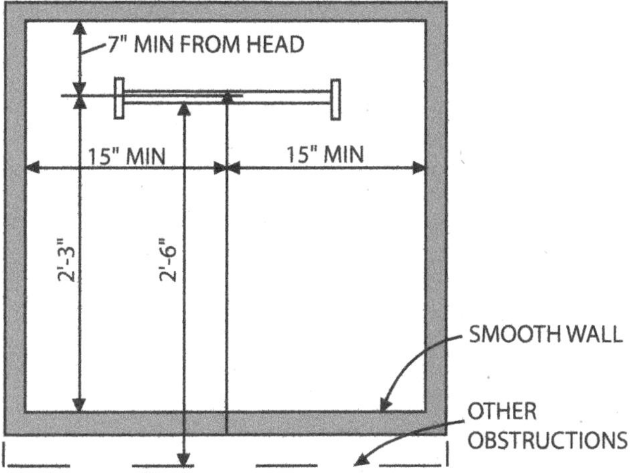

Fire is a danger that in modern times is better controlled than ever before. Fire prevention measures are taught in elementary schools. Fire retardant materials are becoming more common in the building and construction trades. Fire suppression systems are now more sophisticated and effective than ever. Many of these advances have been due to the unceasing efforts of the National Fire Protection Association (NFPA) and other organizations to understand fire and its properties.

- *Pyrophoric* materials ignite spontaneously when in contact with air.
- *Hypergolic* is used to describe a violent reaction that occurs when two materials come in contact with each other.
- *Pyrolysis* is the process of decomposition in the presence of heat.
- *Deflagration* is a rapidly burning fire that produces a flame speed that is slower than the speed of sound.

According to the *Fire Protection Handbook*,® in order for combustion to occur, four components are necessary: oxygen (an oxidizing agent); fuel (a substrate); heat (ignition); and a self-sustained chemical reaction (also referred to as a chain reaction). These components can be vividly described as the "fire tetrahedron." All four components of the tetrahedron must be in place for combustion to occur. Remove any one of the components, and combustion will not occur. Even if ignition has already occurred, the fire will be extinguished when one of the other components is removed from the reaction.

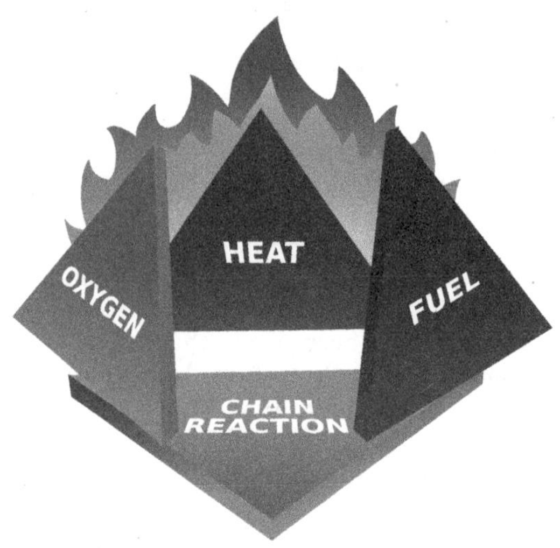

Gustavb/Wikimedia Commons/Public Domain

The NFPA also classifies fire extinguishers according to their hazard use and provides guidelines on egress, spacing of extinguishers, testing of extinguishers, symbolism used on extinguisher canisters, and extinguishing media.

NFPA 704 Symbol

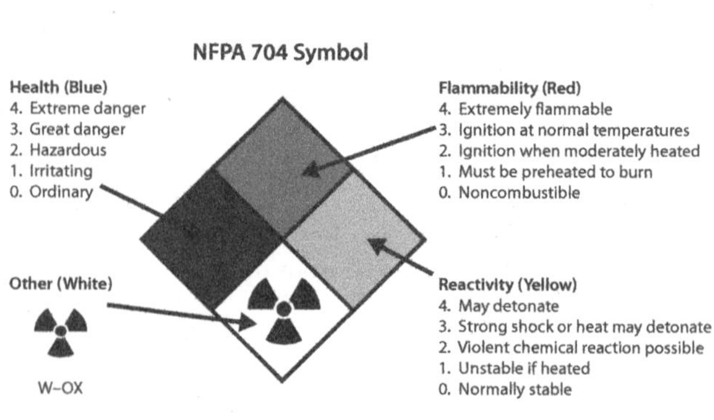

CLASSIFICATIONS OF MATERIALS

The NFPA has ranked and subclassified different types of flammable and combustible materials based on each material's flash point and boiling point. The lower the material's flash point, the more flammable the material is. The higher the material's flash point, the more difficult it becomes to ignite the material to the point of its becoming combustible.

Flammable ≤ 100°F Flash Point
Vapor Pressure < 40 psia at 100°F

Class I	Flash point < 100°F	Vapor pressure ≤ 40 psia at 100°F
Class IA	Flash point < 73°F	Boiling point < 100°F
Class IB	Flash point < 73°F	Boiling point ≥ 100°F
Class IC	Flash point ≥ 73°F, < 100°F	

Combustibles ≥ 100°F, < 200°F Flash Point

Class II	Flash point ≥ 100°F, below 140°F
Class III	Flash point ≥ 140°F
Class IIIA	Flash point ≥ 140°F, below 200°F
Class IIIB	Flash point ≥ 200°F

Reference: After NFPA 30, NSC

Flammable and Combustible Liquid Classification Comparison

Flash Point Closed Cup	<20°F (−7°C)	20°F (−7°C)– 100°F (38°C)	100°F (38°C)–1 40°F (60°C)	140°F (60°C)– 150°F (66°C)	150°F (66°C)– 200°F (93°C)
OSHA	Flammable	Flammable	Combustible	Combustible	Combustible
ANSI	Extremely Flammable	Flammable	Flammable <141°F (60.5°C)	Combustible	Combustible
RCRA (EPA)	Ignitable	Ignitable	Ignitable		
DOT	Flammable	Flammable	Flammable <141°F (60.5°C)	Combustible	Combustible
CPSC	Extremely Flammable	Flammable	Combustible	Combustible	
NFPA 30	Class I	Class I	Class II	Class III	Class III

GHS Flammable and Combustible Liquid Criteria

Criteria	GHS Category	Transport Class/Packing Group
Flash Point <73°F (23°C) Initial boiling point ≤95°F (35°C)	1	3, I
Flash Point <73°F (23°C) Initial boiling point >95°F (35°C)	2	3, II
Flash Point ≥73°F (23°C) and ≤140°F (60.5°C)	3	3, III
Flash Point >140°F (60.5°C) and ≤199.4°F (93°C)	4	Combustible Liquid, PG III [DOT uses <200°F (93°C)]

DEFINITIONS

- Flame spread rating—burning characteristics of material
- Fire loading—maximum amount of heat generated, including from furniture and drapes
- Fireproofing—insulating from heat
- Fire doors—hourly rating from NFPA 80

EGRESS, TRAVEL DISTANCE TO

- 200 ft for industrial structure
- 250 ft with fully automatic sprinklers
- 400 ft for one-story, low-hazard structure with sprinklers
- 75 ft for high-hazard occupancy with two exits

Fire Alarms

A—manual	Through operator to emergency response unit
B—automatic	Directly to emergency response unit, pull stations needed
Central station	Vendor provides remote monitoring service
Local	Building alarm only

Fire Detectors—Thermal

Fixed temperature thermal	Two metals with different coefficients of expansion, expanding components close circuit
Pneumatic	Nonelectrical system, mechanical, change in air pressure engages system
Rate compensation	Reacts at fixed temperature
Rate of rise	Reacts at selected rate of temperature change, 12°F typical

Fire Extinguishers—Classifications

Type	Works on	Mnemonic Device	Symbol
Class A	Ordinary combustibles	A for ash	Green triangle
Class B	Flammable liquids, gases, grease	B for barrel	Red square
Class C	Electrical	C for current	Blue circle
Class D	Metals—magnesium, titanium, sodium, potassium		Yellow star

Types of Extinguishers

	Test interval (years)
Soda acid (soldered brass shells) (until 1/1/82)	(1)
Soda acid (stainless steel shell)	5
Cartridge operated water and/or antifreeze	5
Stored pressure water and/or antifreeze	5
Wetting agent	5
Foam (soldered brass shells) (until 1/1/82)	(1)
Foam (stainless steel shell)	5
Aqueous film forming foam (AFFF)	5
Loaded stream	5
Dry chemical with stainless steel	5
Carbon dioxide	5
Dry chemical, stored pressure, with mild steel, brazed brass or aluminum shells	12
Dry chemical, cartridge or cylinder operated, with mild steel shells	12
Halon 1211	12
Halon 1301	12
Dry powder, cartridge or cylinder operated with mild steel shells	12

FIRE EXTINGUISHERS—DRY CHEMICAL EXTINGUISHER TYPES

- Sodium bicarbonate
- Potassium bicarbonate
- Mono-ammonium phosphate
- Potassium chloride
- Urea-potassium chloride

FIRE EXTINGUISHERS—FOAM EXTINGUISHING MEDIA TYPES

- Aqueous film forming foam (AFFF)
- Fluoroprotein
- Protein

HYDROSTATIC TESTING SCHEDULE FOR EXTINGUISHER CYLINDERS

- 12 years—dry chemical, halon, dry powder
- 5 years—all except dry chemical in stainless steel shell

Smoke Detectors

Flame	
Ionization	Resistance of heated element changes, galvanometer, smoke aerosols stick to ions needing more voltage
Infrared	Susceptible to sunlight, very fast response
Photoelectric	Change in amount of light reaching detector, particles obscure light beams, better for large rooms because does not dissipate as fast as heat
Ultraviolet	Insensitive to sunlight or artificial lighting

SPRINKLER SYSTEMS

Sprinkler systems are a specialty of their own. Sprinkler head standardization and coverage recommendations ensure adequate water spread. Sprinkler system types and detector activation characteristics allow customization for any space.

Sprinkler System Caps

Combination dry pipe and pre-action	Air in pipes, detector opens valve
Deluge	Heads are open at all times but pipes are empty, detector sends water
Dry pipe	Air and nitrogen under pressure, sprinkler opens, pressure drops, water flows
Pre-action automatic	Water present at all times in system, auxiliary system, detector activated
Wet pipe	Pressurized water, sprinkler opens, water flows

Sprinkler System Colors	°C	°F	Type
Uncolored	38	100	Ordinary
White	66	150	Intermediate
Blue	107	225	High
Red	150	300	Extra high
Green	190	375	Very extra high
Orange	218	425	Ultra high
Orange	246	475	Ultra high

STANDARDS

National Fire Protection Association publications offer general and specific guidelines on many subjects related to fire safety. The more commonly used publications are listed here.

Standards—NEC Article 500-503 Hazardous Locations

Class	Division 1	Division 2
I Flammable gases, vapors, and liquids	Hazard present in ignitable/explosive concentrations	Not normally in explosive concentrations
II Dusts	Ignitable quantities present	Not normally in ignitable concentrations
III Fibers and fine particles	Material actively handled or used in manufacturing	Storage or handled in storage

STANDARDS—NFPA 13

- 1 sprinkler/riser for 40,000 ft^2 warehouse
- 1 sprinkler/riser for 52,500 ft^2 office
- 0.1 gpm/ft^2—rate for subsurface foam
- 1 head covers 10×10 ft $= 100$ ft^2
- Most losses from closed water supply valves

STANDARDS—NFPA 30

- Maximum in flammable storage cabinets—60 gal flammables, 120 gal combustibles, 20B extinguisher
- 25 gal maximum outside storage cabinet
- 3 cabinets in one room
- 1,100 gal maximum in one area
- Storage areas 20 ft from building
- Halon naming system—carbons/fluorides/chlorides/bromides/iodine

STANDARDS—NFPA 101

- Means of egress, access to exits, exit, and exit discharge
- Minimum exit width for existing structures = 28 in.

- Minimum exit width for new buildings = 36 in.
- Minimum exit width needed for wheelchair (per ADA) = 32 in.

STANDARDS—NFPA FIRE SAFETY

NFPA 10—Standard for Portable Fire Extinguishers
NFPA 13—Installation of Sprinkler Systems
NFPA 30—Flammable and Combustible Liquids Code
NFPA 68—Guide for Venting of Deflagrations
NFPA 69—Standard on Explosion Prevention Systems
NFPA 70—National Electrical Code
NFPA 77—Recommended Practice on Static Electricity
NFPA 80—Standard for Fire Doors and Fire Windows
NFPA 101—Life Safety Code
NFPA 220—Standard of Types of Building Construction
NFPA 251—Standard Methods of Fire Tests of Fire Endurance of Building Construction and Materials
NFPA 495—Explosive Materials Code
NFPA 704—Standard for the Identification of Fire Hazards of Materials for Emergency Response

OSHA and the NFPA have similar guidelines regarding the maximum amount of flammable liquids in one location per hazard classification. The following table contains a cross-referenced listing of current national consensus standards that contain information and guidelines that would be considered acceptable for complying with requirements in the specific sections of OSHA *CFR* 1910, subpart L.

OSHA 1910 Subpart L	National Consensus Standard
1910.156	ANSI/NFPA No. 1972, Structural Fire Fighters Helmets ANSI Z88.5, American National Standard, Practice for Respirator Protection for the Fire Service ANSI/NFPA No. 1971, Protective Clothing for Structural Fire Fighters NFPA No. 1041, Fire Service Instructor Professional Qualifications
1910.157	ANSI/NFPA No. 10, Portable Fire Extinguishers
1910.158	ANSI/NFPA No. 18, Wetting Agents ANSI/NFPA No. 20, Centrifugal Fire Pumps NFPA No. 21, Steam Fire Pumps ANSI/NFPA No. 22, Water Tanks NFPA No. 24, Outside Protection NFPA No. 26, Supervision of Valves NFPA No. 13E, Fire Department Operations in Properties Protected by Sprinkler, Standpipe Systems ANSI/NFPA No. 194, Fire Hose Connections NFPA No. 197, Initial Fire Attack, Training for NFPA No. 1231, Water Supplies for Suburban and Rural Fire Fighting
1910.159	ANSI/NFPA No. 13, Sprinkler Systems NFPA No. 13A, Sprinkler Systems, Maintenance ANSI/NFPA No. 18, Wetting Agents ANSI/NFPA No. 20, Centrifugal Fire Pumps ANSI/NFPA No. 22, Water Tanks NFPA No. 24, Outside Protection NFPA No. 26, Supervision of Valves ANSI/NFPA No. 72B, Auxiliary Signaling Systems NFPA No. 1231, Water Supplies for Suburban and Rural Fire Fighting
1910.160	ANSI/NFPA No. 11, Foam Systems ANSI/NFPA No. 11A, High Expansion Foam Extinguishing Systems ANSI/NFPA No. 11B, Synthetic Foam and Combined Agent Systems ANSI/NFPA No. 12, Carbon Dioxide Systems ANSI/NFPA No. 12A, Halon 1301 Systems ANSI/NFPA No. 12B, Halon 1211 Systems ANSI/NFPA No. 15, Water Spray Systems ANSI/NFPA No. 16, Foam-Water Spray Systems ANSI/NFPA No. 17, Dry Chemical Systems ANSI/NFPA No. 69, Explosion Suppression Systems

1910.161	ANSI/NFPA No. 11B, Synthetic Foam and Combined Agent Systems ANSI/NFPA No. 17, Dry Chemical Systems
1910.162	ANSI/NFPA No. 12, Carbon Dioxide Systems ANSI/NFPA No. 12A, Halon 1211 Systems ANSI/NFPA No. 12B, Halon 1301 Systems ANSI/NFPA No. 69, Explosion Suppression Systems
1910.163	ANSI/NFPA No. 11, Foam Extinguishing Systems ANSI/NFPA No. 11A, High Expansion Foam Extinguishing Systems ANSI/NFPA No. 11B, Synthetic Foam and Combined Agent Systems ANSI/NFPA No. 15, Water Spray Fixed Systems ANSI/NFPA No. 16, Foam-Water Spray Systems ANSI/NFPA No. 18, Wetting Agents NFPA No. 26, Supervision of Valves
1910.164	ANSI/NFPA No. 71, Central Station Signaling Systems ANSI/NFPA No. 72A, Local Protective Signaling Systems ANSI/NFPA No. 72B, Auxiliary Signaling Systems ANSI/NFPA No. 72D, Proprietary Protective Signaling Systems ANSI/NFPA No. 72E, Automatic Fire Detectors ANSI/NFPA No. 101, Life Safety Code
1910.165	ANSI/NFPA No. 71, Central Station Signaling System ANSI/NFPA No. 72A, Local Protective Signaling Systems ANSI/NFPA No. 72B, Auxiliary Protective Signaling Systems ANSI/NFPA No. 72C, Remote Station Protective Signaling Systems ANSI/NFPA No. 72D, Proprietary Protective Signaling Systems ANSI/NFPA No. 101, Life Safety Code
Metric Conversion	ANSI/ASTM No. E380, American National Standard for Metric Practice

STANDPIPES
Classes of Standpipes

Standpipes	Size	Used By
Class I	2½ in. hose connection	Fire department
Class II	1½ in. hose connection	Building occupants on incipient fires
Class III	Both types of hose connections	Trained building occupants on more advanced fires

STORAGE

Because of the different properties of flammables and combustibles, the NFPA and OSHA issue guidelines on the amounts that can be stored in one place, on the type of location, and on the size of the storage container.

VOLATILITY DUE TO TEMPERATURE/ PRESSURE CHANGES

An increase in temperature causes an increase in vapor pressure (VP). A material with a high vapor pressure is highly volatile. If the VP is less than 1.0, the substance will rise, and if the VP is greater than 1.0, the substance will sink. The higher the vapor pressure, the faster the substance evaporates.

LFL × 100 = VP/14.7

    VP = vapor pressure in absolute at flash point temperature
    LFL = lower flammable limit

A decrease in temperature or pressure causes the LFL to increase and the upper flammable limit (UFL) to decrease, decreasing the flammable range. Similarly, an

Maximum Allowable Size of Containers and Portable Tanks

Type	Class IA	Class IB	Class IC	Class II	Class III
	Flammable Liquids			Combustibles	
Glass/plastic	1 pt	1 qt	1 gal	1 gal	1 gal
Metal (gal) (non-DOT)	1	5	5	5	5
Safety cans (gal)	2	5	5	5	5
Metal drums (gal) (DOT spec)	60	60	60	60	60
Approved portable tanks (gal)	660	660	660	660	660

Reference: OSHA 1910.106

increase in temperature causes the LFL to decrease and the UFL to increase, increasing the flammable range and the danger. A combustion meter reads both the percent of LFL and the lower explosive limit (LEL). If the specific gravity is less than 1.0, the substance will float on water.

WATER CALCULATIONS

- Force = pressure × area
- Pressure = head × density
- Discharge velocity = $\sqrt{(2 \times \text{acceleration of gravity} \times \text{head})}$
- Rate of flow (Q) = area of cross section × velocity at cross section

EXPLOSIONS

Explosive materials are a special class of hazardous materials. Explosions, by their very nature, are an instantaneous phenomenon but may create cascading reactions due to fire and shock waves. Explosion severity is a function of maximum explosion pressure and maximum rate of pressure rise. Ignition sensitivity is a function of ignition temperature, minimum energy of ignition, and minimum concentration. Explosion hazards of dusts are rated on a scale using Pittsburgh coal dust as the standard. Dust explosion intensity is determined by particle concentration and particle size. Where explosives are present, there must be no smoking within 50 ft.

Blast Wave Effects due to Overpressure

Overpressure (psi)	Effect
0.5–1.0	Shatter glass
1.0	Knock down a person
2.0–3.0	Shatter 8–12 in. block concrete wall
5.0	Eardrum rupture threshold, break utility pole
7.0	Overturn loaded railcar
11.0	Lung damage threshold
15.0	50% of eardrums ruptured

HazMat Classification—DOT

Class	Division	Category
1		Explosives
	1.1	Mass explosion hazard
	1.2	Projection hazard
	1.3	Fire hazard and minor blast hazard and/or minor projection hazard
	1.4	Minor blast hazard
	1.5	Very insensitive explosives
	1.6	Extremely insensitive detonating substance
2		Gases
	2.1	Flammable gases
	2.2	Compressed nonflammable gases
	2.3	Poisonous gases
3		Flammable and combustible liquids
4		Flammable solids
	4.1	Flammable solids
	4.2	Spontaneously combustible
	4.3	Dangerous when wet
5		Oxidizers and organic peroxides
	5.1	Oxidizers
	5.2	Organic peroxides
6		Poisons
	6.1	Poisonous
	6.2	Infectious substance
7		Radioactive
8		Corrosive
9		Miscellaneous
ORM-D		Other regulated materials (consumer commodities)

Reference: DOT 49 CFR 172.407.

Lower Explosive Limit of Some Commonly Used Solvents

Solvent	Cubic ft per gallon of vapor of liquid at 70°F	Lower explosive limit in % by volume of air at 70°F
Acetone	44.0	2.6
Amyl Acetate (iso)	21.6	1.0[1]
Amyl Alcohol (n)	29.6	1.2
Amyl Alcohol (iso)	29.6	1.2
Benzene	36.8	1.4[1]
Butyl Acetate (n)	24.8	1.7
Butyl Alcohol	35.2	1.4
Butyl Cellosolve	24.8	1.1
Cellosolve	33.6	1.8
Cellosolve Acetate	23.2	1.7
Cyclohexanone	31.2	1.1[1]
1,1 Dichloroethylene	42.4	5.9
1,2 Dichloroethylene	42.4	9.7
Ethyl Acetate	32.8	2.5
Ethyl Alcohol	55.2	4.3
Ethyl Lactate	28.0	1.5[1]
Methyl Acetate	40.0	3.1
Methyl Alcohol	80.8	7.3
Methyl Cellosolve	40.8	2.5
Methyl Ethyl Ketone	36.0	1.8
Methyl n-Propyl Ketone	30.4	1.5
Naphtha (VM&P) (76° Naphtha)	22.4	0.9
Naphtha (100° flash) Safety Solvent— Stoddard Solvent	23.2	1.0
Propyl Acetate (n)	27.2	2.8
Propyl Acetate (iso)	28.0	1.1
Propyl Alcohol (n)	44.8	2.1
Propyl Alcohol (iso)	44.0	2.0
Toluene	30.4	1.4
Turpentine	20.8	0.8
Xylene (o)	26.4	1.0

		Sample Supervisor's Safety Checklist—Fire Prevention
Supervisor:		
Date:		**Department/Area:**
Yes	**No**	**Checklist Area**
Area Safety—check that:		
		all flammable liquids are stored in approved, vented storage lockers
		oily rags are disposed of in approved containers
		electrical motors are checked daily for overheating
		trash does not accumulate in work areas
		flammable material is not stored under stairs
		no storage is allowed in electrical utility rooms
		waste dumpsters are not located next to buildings
		smoking is allowed only in designated areas—waste containers provided
		flammable storage areas have No Smoking signs posted
		combustible gases are properly stored
		bonding and grounding of flammable liquid containers is enforced
		incompatible chemicals are stored in separate areas
		Hot Work permits are used for all welding and spark-producing operations
		access to utility and equipment rooms is restricted to authorized employees
		combustible and flammable material is not stored near gas-fired equipment
Employee Training—workers are trained to:		
		recognize and report fire hazards
		activate fire alarms
		understand proper storage of flammable liquids
		(supervisors) properly use fire extinguishers

Knowledge of basic chemical principles regarding the atom, molecule, and compound is necessary to understand how solids, liquids, and gases act in the environment. From this, the pH scale, basic gas laws, and constant formulas provide the mathematical basis for calculations.

AGENCIES AND REGULATIONS

Many federal agencies and regulations have some control over safety, either for the public or for workers. They are nearly always referred to by their initials.

CAA (Clean Air Act)—air stack emissions

CERCLA (Comprehensive Environmental Responsibility, Compensation, and Liability Act)—potentially responsible parties (PRPs), abandoned hazardous waste sites.[8]

CPSC (Consumer Product Safety Commission)—protects consumers, can order manufacturer to recall a product if it has a substantial risk of injury or death. Its five members are appointed by the U.S. president and approved by Congress, but they are independent and not part of a cabinet department. May impose a $500,000 maximum penalty.

CWA (Clean Water Act)—stream effluents

DOT (Department of Transportation)—cabinet-level department, trucking, Federal Highway Administration (roadways, traffic control devices, pedestrian safety), Federal Railroad Administration, Coast Guard

EPA (Environmental Protection Act)—regulated under TSCA to assess chemical risks, regulate new chemicals, give 90-day notice; regulated under RCRA to monitor hazardous wastes

FAA (Federal Aviation Administration)—subagency of DOT

FIFRA (Federal Insecticide, Fungicide, and Rodenticide Act)

NEISS (National Electronic Injury Surveillance System)—database of hospital records maintained by the CPSC

NEPA (National Environmental Policy Act)

NHTSA (National Highway and Traffic Safety Administration)—reduce accidents, promote air bags and seat belts

NIH (National Institutes of Health)—NIOSH and CDC under its umbrella

NPDES (National Pollution Discharge Elimination System)—effluent discharge

NRC (National Response Center)

NTSB (National Transportation Safety Board)—investigates accidents, reports to president

RCRA (Resource Conservation and Recovery Act)—authorizes EPA to regulate activities related to waste management. Ten subtitles (A–J). Subtitle C covers hazardous wastes from cradle to grave, with criteria for determining which solid waste is hazardous and sets forth regulations for generators, transporters, and treatment, storage, and disposal facilities (TSDFs). Subtitle D regulates solid waste disposal facilities and municipal solid waste land facilities. Local municipalities are the lead agencies for non-hazardous solid waste. Subtitle I regulates hazardous substances and petroleum products held in underground storage tanks (USTs) with requirements for tank standards and release cleanup procedures.

SARA (Superfund Amendments and Reauthorization Act; subpart of CERCLA)—Community Right-to-Know provisions (Title III), abandoned waste sites (Titles I and II)

SPCC (Spill Prevention Control Countermeasures)

TCLP (Toxicity Characteristic Leaching Procedure)

TSCA (Toxic Substance Control Act)—90-day pre-manufacturing notice, regulates new chemical products, chlorinated fluorocarbons (CFCs), asbestos, PCBs, dioxin

UST—underground storage tank

Reporting Requirements	CWA Section 311	CERCLA Section 103	SARA Title III Section 304	RCRA Subtitle I (UST Program)
Type of release subject to reporting	Discharge to surface water	Releases into the environment (outside enclosed building or structure)	Releases into the environment with potential to result in exposure to persons off-site	Ground water, surface water, or surface soils
Hazardous substances subject to reporting	Oil and hazardous substances listed in 40 *CFR* 116.4	CERCLA hazardous substances listed in 40 *CFR* 302.4	Extremely hazardous substances listed in 40 *CFR* 355.20 and CERCLA hazardous substances	Petroleum and CERCLA hazardous substances (excluding RCRA hazardous wastes)
Trigger amount for reporting	All discharges of oil that form a sheen and discharges of hazardous substances that equal or exceed the CERCLA RQ	Releases that equal or exceed the CERCLA RQ	Releases that equal or exceed the CERCLA RQ, or one pound for EHSs that are not CERCLA hazardous substances	Any suspected or confirmed release, spills that equal or exceed the CERCLA RQ, and spills of petroleum that exceed 25 gallons or that result in a sheen
Facilities subject to reporting	All facilities and vessels	All facilities and vessels, except where consumer products are in consumer use	All facilities that produce, use, or store an OSHA hazardous chemical	Facilities with an underground storage tank system, the volume of which is 10% or more beneath the surface of the ground

Reporting Requirements	CWA Section 311	CERCLA Section 103	SARA Title III Section 304	RCRA Subtitle I (UST Program)
Parties responsible for reporting	Person in charge	Person in charge	Owner/operator	Owner/operator
When report is required	Immediately	Immediately	Immediately, with written follow-up report	Within 24 hours
To whom to report	National Response Center	National Response Center	Relevant SERCs and LEPCs; for transportation-related releases, the 911 emergency telephone number or the operator	Implementing agency (usually a state agency)
Maximum penalties	$10,000 and/or prison sentence of one year	$50,000 and/or prison sentence of three years	$50,000 and/or prison sentence of three years	$25,000 per day for each day of noncompliance

Regulatory Aspect	CESQG 40 CFR 261.5	SQG 40 CFR 62.34 (d)&(f)	LQG 40 CFR 262.34(a–c)
Generation Limits (kg/ month)			
Acute Hazardous Waste	≤ 1	≤ 1	> 1
Acute Hazardous Waste Spill Residue	≤ 100	–	–
Total Hazardous Waste	≤ 100	100–1,000	> 1,000
Maximum Storage Limit (kg)			
Acute Hazardous Waste	1		
Acute Hazardous Waste Spill Residue	100		
Total Hazardous Waste	1,000		
General Extent of Regulation	Exempt from Subtitle C except 40 *CFR* 262.11 (hazardous waste determination)	Reduced generator accumulation requirements	Full Subtitle C regulation

Potential Incompatibilities of Common Chemical Classes

Class	Incompatible Class	Potential Result of Reaction
acid	base (caustic), water	heat, violent reaction
	alcohols, electropositive metals (not Al or Mg), organic compounds containing unsaturations or O, NO_2, or NH_2 functional groups	heat, violent reaction including possible generation of flammable gases and fire
	cyanides, sulfides	generation of toxic HCN, H_2S
	acid chlorides	heat, violent reaction, generation of toxic HCl gas
base (caustic)	electropositive metals (not Fe), reactive metals and metal hydrides	heat, violent reaction including possible generation of flammable gases and fire
	acids, acid chlorides, water, alcohols, organic compounds containing unsaturations or O, NO_2, or NH_2 functional groups	heat, violent reaction
alcohols, water	acids, bases (caustics)	heat, violent reaction
	acid chlorides	heat, violent reaction, generation of toxic HCl gas
reactive metals, metal hydrides	water, oxidizers, organic compounds containing OH, NH_2, or halogens	heat, violent reaction, possibly including fire or explosion
oxidizers	acids, reactive metals, metal hydrides, electropositive metals (especially if finely divided), organic compounds containing unsaturations or O, NO_2, or NH_2 functional groups	heat, violent reaction, possibly including fire or explosion
	acid chloride	heat, violent reaction, generation of toxic HCl gas

Common Classes of Potentially Reactive Chemicals

Chemical Class	Includes	Examples
acids	strong mineral acids and their solutions plus some concentrated weak acids	sulfuric acid, hydrochloric acid, glacial acetic acid, concentrated phosphoric acid
bases (caustics)	alkali metal hydroxides and their solutions plus some concentrated weak bases	sodium and potassium hydroxide, concentrated calcium hydroxide
alcohols	hydroxyalkanes	ethanol, methanol, 1-octanol
reactive metals	strongly electropositive metals	sodium, potassium, lithium, calcium
electropositive metals	metals that readily oxidize—at least with halogens or on the surface—but that are attacked only slowly (if at all) by water at room temperature	iron, aluminum, magnesium, zinc
metal hydrides	binary compounds of metals with hydrogen	sodium hydride, lithium hydride
strong oxidizers	DOT oxidizers that react strongly with organic compounds	perchlorates, chlorates, chlorites, hypochlorites, peroxides, permanganates, some nitrates, some acids (nitric, perchloric, chloric)
halogenated compounds	organic compounds with one or more halogens (F, Cl, Br, I) (Fluorinated compounds tend to have low reactivities compared to the other materials)	trichloroethylene, methylene chloride, bromoform, methyl iodide
acid chlorides	a special group of halogenated compounds in which an OH on an acid is replaced by a Cl	acetyl chloride, benzoyl chloride, sulfuryl chloride, thionyl chloride, phosgene, phosphorus trichloride

Alcohol Name	Flash Point	LEL–UEL	Boiling Point	Vapor Density
Methanol	52	6.0–36%	147°F	1.1
Ethanol (96%)	55	3.3–19%	173°F	1.6
Isopropanol	53	2.0–17%	181°F	2.1
Ethylene glycol	232	3.2–??%	387°F	2.0

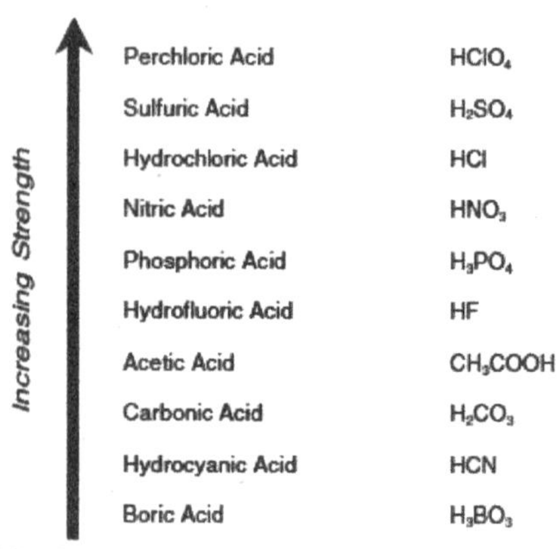

OSHA HAZARD COMMUNICATION STANDARD (HAZCOM) 1910.1200

Hazard Communication Safety Data Sheets

The Hazard Communication Standard (HCS) requires chemical manufacturers, distributors, and importers to provide Safety Data Sheets (SDSs—formerly known as Material Safety Data Sheets, or MSDSs) to disclose the dangers of hazardous chemical products. As of June 1, 2015, SDSs will be in a uniform format and include the section numbers, the headings, and associated information under the headings below:

Section 1, Identification—includes product identifier; manufacturer or distributor name, address, and phone number; emergency phone number; recommended use; restrictions on use.

Section 2, Hazard(s) identification—includes all hazards regarding the chemical; required label elements.

Section 3, Composition/information on ingredients—includes information on chemical ingredients; trade secret claims.

Section 4, First-aid measures—includes important symptoms/effects, acute, delayed; required treatment.

Section 5, Fire-fighting measures—lists suitable extinguishing techniques, equipment; chemical hazards from fire.

Section 6, Accidental release measures—lists emergency procedures; protective equipment; proper methods of containment and cleanup.

Section 7, Handling and storage—lists precautions for safe handling and storage, including incompatibilities.

Section 8, Exposure controls/personal protection—lists OSHA's Permissible Exposure Limits (PELs); Threshold Limit Values (TLVs); appropriate engineering controls; personal protective equipment (PPE).

Section 9, Physical and chemical properties—lists the chemical's characteristics.

Section 10, Stability and reactivity—lists chemical stability and possibility of hazardous reactions.

Section 11, Toxicological information—includes routes of exposure; related symptoms, acute and chronic effects; numerical measures of toxicity.

Section 12, Ecological information*

Section 13, Disposal considerations*

Section 14, Transport information*

Section 15, Regulatory information*

Section 16, Other information—includes the date of preparation or last revision.

*Note: Since other agencies regulate this information, OSHA will not be enforcing Sections 12 through 15 (29 *CFR* 1910.1200(g)(2)).

Employers must ensure that SDSs are readily accessible to employees.

See Appendix D of *CFR* 1910.1200 for a detailed description of SDS contents.

HAZARD COMMUNICATION STANDARD LABELS

Under its Hazard Communication Standard (HCS), OSHA has updated the requirements for labeling hazardous chemicals. All labels are now required to have pictograms, a signal word, hazard and precautionary statements, the product identifier, and supplier identification. Supplemental information can also be provided on the label as needed.

The Hazard Communication Standard requires pictograms on labels to alert users of the chemical hazards to

Sample Label

Product Identifier
CODE _______________________________
Product Name_______________________________

Supplier Identification
Company Name _______________________________
Street Address _______________________________
City _______________________ State_______________
Postal Code _______________ Country _______________
Emergency Phone Number _______________________

Hazard Pictograms

Signal Word
Danger

Precautionary Statements

Keep container tightly closed. Store in cool, well-ventilated place that is locked.
Keep away from heat/sparks/open flame. No smoking.
Only use nonsparking tools.
Use explosion-proof electrical equipment.
Take precautionary measure against static discharge.
Ground and bond container and receiving equipment.
Do not breathe vapors.
Wear protective gloves.
Do not eat, drink, or smoke when using this product.
Wash hands thoroughly after handling.
Dispose of in accordance with local, regional, national, and international regulations as specified.

In Case of Fire: Use dry chemical (BC) or carbon dioxide (CO_2) fire extinguisher to extinguish.

First Aid
If exposed, call the nearest Poison Center.
If on skin (on hair): Take off immediately any contaminated clothing.
Rinse skin with water.

Hazard Statement

Highly flammable liquid and vapor.
May cause liver and kidney damage.

Supplemental Information
Directions for use

____
____
____

Fill Weight:_______________ Lot Number: _______________
Gross Weight: _______________ Fill Date: _______________
Expiration Date: _______________________________

which they may be exposed. Each pictogram consists of a symbol on a white background, framed within a red border, and representing a distinct hazard(s). The pictogram on the label is determined by the chemical hazard classification.

HCS Pictograms and Hazards

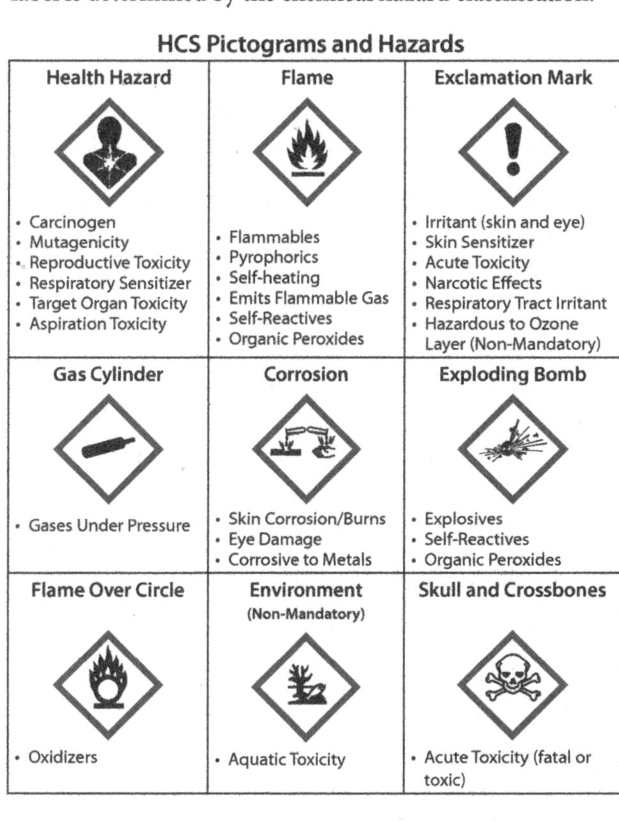

Health Hazard	Flame	Exclamation Mark
• Carcinogen • Mutagenicity • Reproductive Toxicity • Respiratory Sensitizer • Target Organ Toxicity • Aspiration Toxicity	• Flammables • Pyrophorics • Self-heating • Emits Flammable Gas • Self-Reactives • Organic Peroxides	• Irritant (skin and eye) • Skin Sensitizer • Acute Toxicity • Narcotic Effects • Respiratory Tract Irritant • Hazardous to Ozone Layer (Non-Mandatory)
Gas Cylinder	Corrosion	Exploding Bomb
• Gases Under Pressure	• Skin Corrosion/Burns • Eye Damage • Corrosive to Metals	• Explosives • Self-Reactives • Organic Peroxides
Flame Over Circle	Environment (Non-Mandatory)	Skull and Crossbones
• Oxidizers	• Aquatic Toxicity	• Acute Toxicity (fatal or toxic)

WORKER TRAINING

Chemicals pose a wide range of health and safety hazards. OSHA's Hazard Communication Standard (29 *CFR* 1910.1200) is designed to ensure that information about these hazards and associated protective measures is communicated to workers.

Worker training must be provided if cleaning chemicals are hazardous. This training must be provided BEFORE the worker begins using the cleaner. Required training under the OSHA Hazard Communication Standard includes:

- Health and physical hazards of the cleaning chemicals;
- Proper handling, use, and storage of all cleaning chemicals being used, including dilution procedures when a cleaning product must be diluted before use;
- Proper procedures to follow when a spill occurs;
- Personal protective equipment required for using the cleaning product, such as gloves, safety goggles, and respirators; and
- How to obtain and use hazard information, including an explanation of labels and SDSs.

The following are important issues to be discussed with workers during training:

- Never mix different cleaning chemicals together. Dangerous gases can be released.
- Cleaning chemicals should not be used to wash hands. Wash hands with water after working with a cleaning chemical, especially before eating, drinking, or smoking.

Employers must provide training to workers at a level and in a language and vocabulary that they can understand.

HAZMAT CLASSIFICATION—DOT

Shipping documents (papers) provide the needed vital information regarding hazardous materials/dangerous goods to initiate protective actions.

Information provided:

- 4-digit identification number, UN or NA
- Proper shipping name
- Hazard class or division number of material
- Packing group

- Emergency response telephone number
- Information describing the hazards of the material (entered on or attached to shipping document)

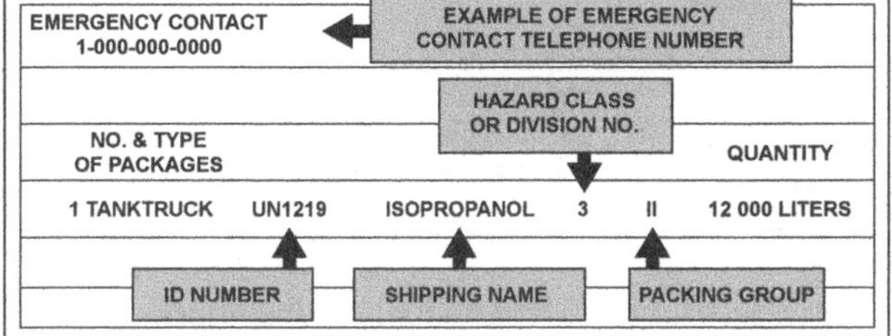

EXAMPLE OF PLACARD AND PANEL WITH ID NUMBER

The 4-digit ID Number may be shown on the diamond-shaped placard or on an adjacent orange panel displayed on the ends and sides of a cargo tank, vehicle or rail car.

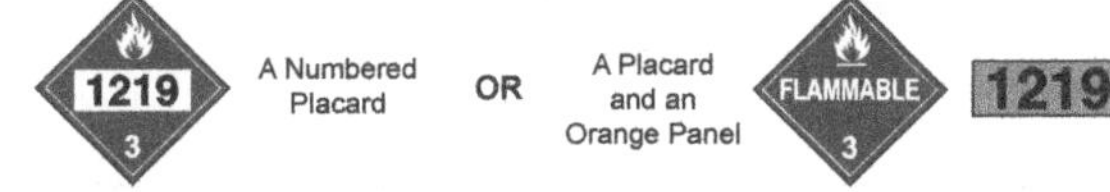

Class 1—Explosives

Division 1.1 Explosives with a mass explosion hazard

Division 1.2 Explosives with a projection hazard

Division 1.3 Explosives with predominantly a fire hazard

Division 1.4 Explosives with no significant blast hazard

Division 1.5 Very insensitive explosives with a mass explosion hazard

Division 1.6 Extremely insensitive articles

Class 2—Gases

Division 2.1 Flammable gases

Division 2.2 Nonflammable, nontoxic* gases

Division 2.3 Toxic* gases

Class 3—Flammable liquids (and Combustible liquids [U.S.])

Class 4—Flammable solids; Spontaneously combustible materials; and Dangerous when wet materials/Water-reactive substances

Division 4.1 Flammable solids
Division 4.2 Spontaneously combustible materials
Division 4.3 Water-reactive substances/
Dangerous when wet materials

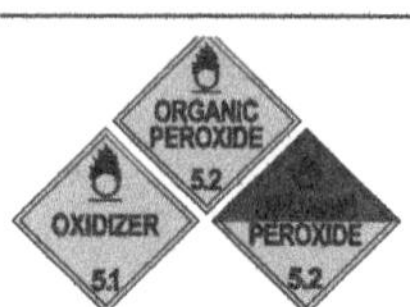

Class 5—Oxidizing substances and Organic peroxides

Division 5.1 Oxidizing substances
Division 5.2 Organic peroxides

Class 6—Toxic* substances and Infectious substances

Division 6.1 Toxic* substances
Division 6.2 Infectious substances

Class 7—Radioactive materials

Class 8—Corrosive substances

Class 9—Miscellaneous hazardous materials/Products, Substances, or Organisms

§172.560

Approach Cautiously from *Upwind, Uphill,* or *Upstream*:
- Stay clear of *Vapor, Fumes, Smoke,* and *Spills.*
- Keep vehicle at a safe distance from the scene.

Secure the Scene:
- Isolate the area and protect yourself and others.

Identify the Hazards Using Any of the Following:
- Placards
- Container labels
- Shipping documents
- Rail Car and Road Trailer Identification Chart
- Safety Data Sheets (SDSs)
- Knowledge of persons on scene
- Applicable guide page

Assess the Situation:
- Is there a fire, a spill, or a leak?
- What are the weather conditions?
- What is the terrain like?
- Who/what is at risk: people, property, or the environment?
- What actions should be taken—evacuation, in-place shelter, or dike?
- What resources (human and equipment) are required?
- What can be done immediately?

Obtain Help:
- Advise your headquarters to notify responsible agencies and call for assistance from qualified personnel.

Respond:
- Enter only when wearing appropriate protective gear.
- Rescue attempts and protecting property must be weighed against you becoming part of the problem.

- Establish a command post and lines of communication.
- Continually reassess the situation and modify the response accordingly.
- Consider the safety of people in the immediate area first, including your own.

HAZARDOUS WASTE OPERATIONS AND EMERGENCY RESPONSE OSHA 1910.120

OSHA's standards on hazardous waste operations and emergency response for general industry and construction industry (29 *CFR* 1910.120 or 29 *CFR* 1926.65) cover all employees involved in:
- Cleanup operations of hazardous substances at uncontrolled hazardous waste sites
- Corrective actions involving cleanup procedures at sites covered by the Resource Conservation and Recovery Act (RCRA)
- Voluntary cleanup operations at sites recognized as uncontrolled hazardous waste sites
- Operations involving hazardous waste that are conducted at treatment, storage, and disposal facilities licensed under the RCRA
- Emergency response operations for hazardous substances

Exceptions are permitted if the employer can demonstrate that the operation does not involve employee exposure or a reasonable possibility of such exposure to hazards.

HAZARDOUS WASTE OPERATIONS

Each employer must have:
- A written, readily accessible safety and health program that identifies, evaluates, and controls safety and health hazards and provides for emergency response.

- A preliminary site evaluation conducted by a qualified person to identify potential site hazards and to aid in the selection of appropriate employee protection methods.
- A site control program to protect employees against hazardous contamination. At a minimum, it must have a site map, site work zones, site communications, safe work practices, the use of a "buddy system," and identification of the nearest medical aid.
- Employee training for everyone working at a hazardous waste site.
- Medical surveillance of workers exposed at or above permissible exposure limits for hazardous substances that is conducted (1) at least annually, (2) when a worker moves to a new worksite, (3) when a worker experiences exposure from unexpected or emergency releases, and (4) at the end of employment.

Other requirements include controls to reduce and monitor exposure levels of hazardous materials, an informational program describing any exposure during operations, and the inspection of drums and containers prior to removal or opening. Decontamination procedures and emergency response plans (described under Emergency Response) must be in place before employees begin working in hazardous waste operations.

Employers must also create safe environments by developing and implementing effective, new technologies.

RCRA SITES

In addition to programs for safety and health, training, medical surveillance, decontamination, new technology, and emergency response, employers at RCRA sites also need the following:

- A written hazard communication program meeting the requirements of 29 *CFR* 1910.1200.
- Procedures to effectively control and handle drums and containers.

Table of Hazardous Waste Regulations

	Purpose	Basic Requirements	Extent of Coverage
RCRA	40 *CFR* 260 to 271 Regulate hazardous waste generators and transporters; manage treatment, storage, and disposal facilities (TSDFs) and cradle to grave regulations.	Permits, technical standards, and groundwater monitoring; corrective action for releases; transportation manifest forms; and UST standards.	Hazardous wastes as defined by EPA; TSDFs; generators of more than 100 kg of hazardous wastes a month.
CERCLA	40 *CFR* 300.61 to 300.71 Locate, assess, and clean potential hazardous waste sites and emergency spills; grant EPS authority to initiate investigations, testing, and monitoring of disposal sites; implement site remediation.	Place a site on the National Priority List because the site poses a threat to human health and the environment.	Removal and response actions; establish strict liability resulting from hazardous wastes; finance response actions through a hazardous-substance-response trust fund; notification requirements for hazardous waste spills and releases; Title III of SARA Community Right-to-Know provisions.
CWA	40 *CFR* 121 to 135, and 403 Protect U.S. waters (surface and groundwater) from direct, indirect, and hazardous point-source pollution discharges.	Establish water-quality standards; set discharge limits.	Restrict pollutant discharges into water; impose monitoring requirements.
CAA	40 *CFR* 69 Protect and enhance U.S. air quality; maintain National Ambient Air Quality Standards (NAAQS); establish National Emissions Standards for Hazardous Air Pollutants (NESHAP).	Prevent air-quality deterioration; set NAAQS and emission standards.	Hazardous air pollutants and VOCs; NAAQS for NO, CO, sulfur dioxide, lead, ozone, and particulate matter.

	Purpose	Basic Requirements	Extent of Coverage
DOT	49 *CFR* 172, 173, 178, 179 Establish standards for the safe transport of hazardous wastes through package and container requirements; establish a labeling system applicable to hazardous waste transportation; document transport to ensure proper treatment and disposal; notification of roadside spills	Requirements for registration, labeling, and vehicle and driver safety; "manifest" documents for shipments of hazardous wastes.	Transportation of hazardous wastes.
OSHA	29 *CFR* Ensure safe and healthy employment conditions; address the proper management of hazardous materials in the workplace by setting standards designed to prevent injury and illness.	Address exposure limits, labeling, protective equipment, control procedures, monitoring, and measuring employee exposure, medical exams, and access to records.	Applicable to facilities with 10 or more workers; OSHA regulations address hazardous wastes including health standards, cancer policy, Hazard Communication Standard, and Hazardous Waste Operation and Emergency Response rules.
TSCA	Screen new chemicals; test new chemicals identified as potential hazards; regulate toxic substance disposal.	Notification of substantial risk by chemical manufacturers who become aware of a chemical threat.	Chemicals that are hazardous, toxic, corrosive, flammable, irritants, or oxidizers.

EMERGENCY RESPONSE

Employers must develop an emergency response plan to handle possible on-site emergencies and coordinate off-site responses. Rehearsed regularly and reviewed/amended periodically, the plan must address personnel roles; lines of authority, training, and communications; emergency recognition and prevention; site security; evacuation routes and procedures; decontamination procedures; emergency medical treatment; and emergency alerting procedures. Training is required before employees engage in hazardous waste operations and emergency response.

INCIDENT COMMAND SYSTEM

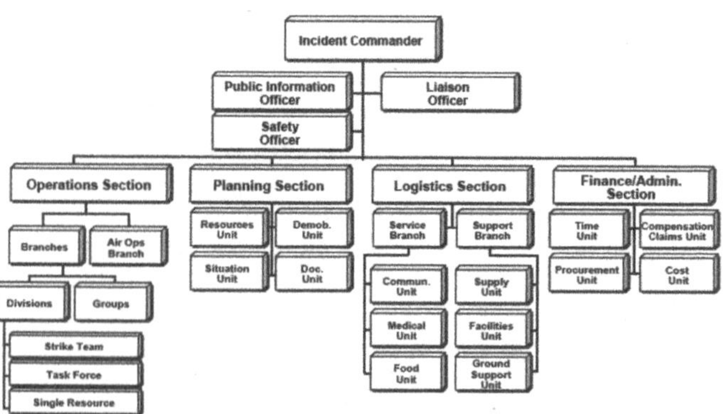

TRAINING REQUIREMENTS

Uncontrolled Hazardous Waste Operations
- 40 hours of initial training; 3 days of actual field experience required for regular employees to be certified

- 24 hours of initial training; 1 day of supervised field experience for employees visiting the site occasionally
- 8 hours of additional waste management training for supervisors and managers
- 8 hours of annual refresher training

Treatment, Storage, and Disposal Facilities Licensed under RCRA

- 24 hours of training
- 8 hours of annual refresher training

Emergency Response Operations at Sites Not RCRA Licensed or at Uncontrolled Hazardous Waste Site Cleanups

(1) **First responders at the "awareness level"** (those who witness or discover a hazardous substance release and initiate the emergency response) must demonstrate competency in areas such as recognizing the presence of hazardous materials in an emergency, the risks involved, and the role they play in their employer's plan.

(2) **First responders at the "operations level"** (those who respond to prevent the spread, exposure to, and further release of hazardous materials) must have **8 hours** of training plus "awareness level" competency.

(3) **Hazardous materials technicians at the "technical level"** (those who respond to stop the release) must have **24 hours** of training equal to that at the "operations level" and know how to implement the employer's plan and carry out decontamination.

(4) **Hazardous materials specialists** (those who require specific knowledge of the substances to be contained) must have **24 hours** of training equal to that at the "technical level" and act as liaison with all government authorities.

(5) **On-scene incident commanders** (those who assume control of the scene) must have **24 hours** of training equal

to that at the "operations level" and demonstrate competence in implementing the incident command system, the employer's plan, and the state and local emergency response plans. Annual refresher training is required for each level of response.

An effective education and training function usually involves establishing and implementing several critical management processes and activities. These include:

- Identifying employee training needs and other competency requirements.
- Selecting effective training methods and techniques.
- Scheduling and providing needed training.

SYSTEMS MODEL FOR DELIVERING TRAINING

Many organizations use a systems approach to providing training and education to employees. This approach involves the following steps:

- Determine education and training requirements.
- Select training methods and develop curriculum.
- Schedule and provide training.
- Evaluate effectiveness and make needed changes.

The ADDIE model is an example of a systematic approach to training and involves:

- Analysis
- Design
- Development
- Implementation
- Evaluation

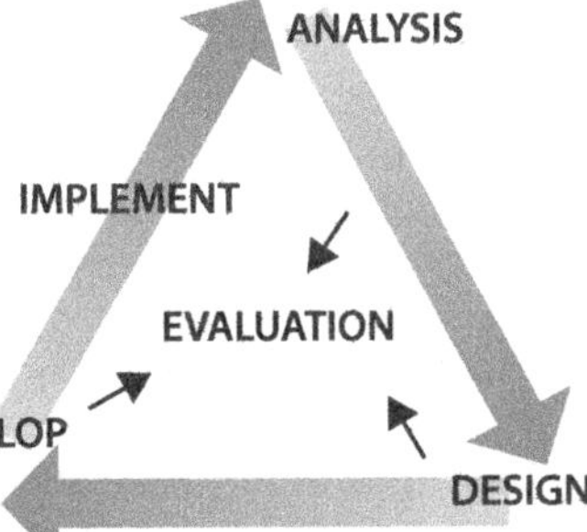

The ADDIE Model
The ADDIE model is a systematic approach to training that is used by instructional designers and learning specialists to determine needs, plan interventions, and evaluate effectiveness.

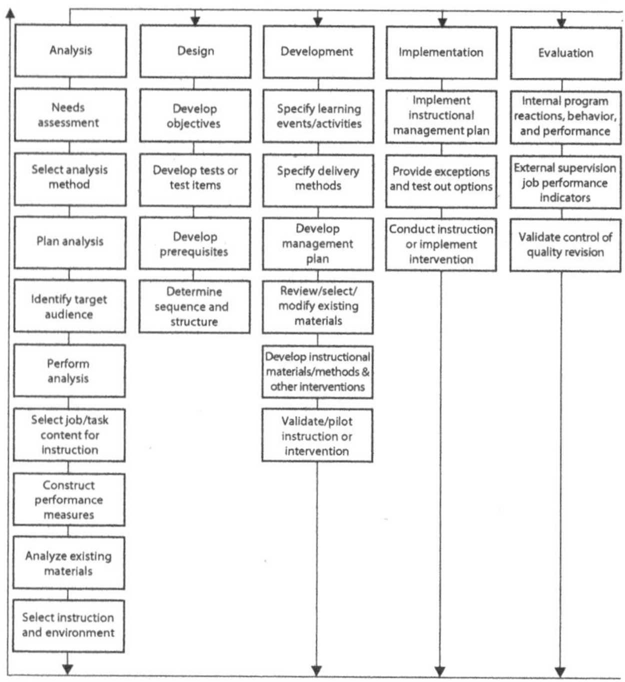

By implementing these processes, programs, and procedures, an organization will be able to ensure that its work force is aware of the organization's safety and health goals, is familiar with their roles and responsibilities, and is able to recognize, control, and mitigate hazards found in the workplace. An effective safety and health education and training function also can play an important role in:
- Reducing occupational injury rates, injury severity, days away from work, and employee health care costs.
- Reducing complaints or incidences that would result in a regulatory enforcement action.
- Reducing employee absenteeism and turnover.
- Decreasing workers' compensation and general liability insurance premiums.

- Increasing employees' skill and motivation, which, in turn, may result in increased productivity and product quality.

Adult learning principles and theories that are integrated into the systematic approach to training models may be found in the ASSE/ANSI Z490.1-2012 standard on *Criteria for Accepted Practices in Safety, Health and Environmental Training*. The following are the basic principles of how adults learn, which are directly applicable to developing safety and health training programs:

Adults are voluntary learners: Most adults learn because they want to. They learn best when they have decided that they need to learn something for a specific reason.

Adults learn needed information quickly: Adults need to see that the subject matter and the methods are relevant to their lives and to what they want to learn. They have a right to know why the information is important to them.

Adults have a good deal of life experience that needs to be acknowledged: They should be encouraged to share their experiences and knowledge.

Adults need to be treated with respect: They resent an instructor who talks down to them or ignores their ideas and concerns.

Adults learn more when they participate in the learning process: Adults need to be involved and actively participate in class.

Adults learn best by doing: Adults need to "try on" and practice what they are learning. They will retain more information when they use and practice their knowledge and skills in class.

Adults need to know where they are heading: Learners need "route maps" with clear objectives. Each new piece of information needs to build logically on the previous piece.

Adults learn best when new information is reinforced and repeated: Adults need to hear things more than once. They need time to master new knowledge, skills, and attitudes. They need to have this mastery reinforced at every opportunity.

Focus Group Brainstorming

Purpose: Identify Possible Solutions

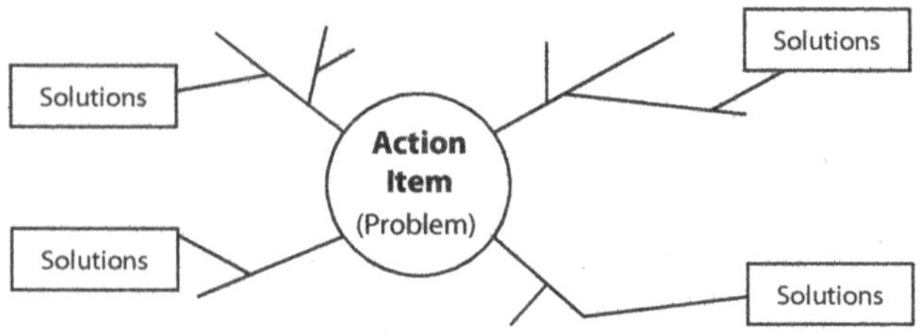

Adults learn best when information is presented in different ways: They learn best when an instructor uses a variety of teaching techniques.

Dr. Malcolm Knowles' theory of andragogy is an attempt to develop a theory specifically for adult learning. Knowles emphasized that adults are self-directed and expect to take responsibility for their decisions. Adult learning programs must therefore accommodate this fundamental characteristic. Andragogy makes the following assumptions about the design of learning:

1. Adults need to know why they need to learn something.
2. Adults need to learn experientially.
3. Adults approach learning as problem solving.
4. Adults learn best when the topic is of immediate value.

In practical terms, andragogy means that instruction for adults needs to focus more on the process and less on the content being taught. Strategies such as case studies, role playing, simulations, and self-evaluation are most useful. In addition, instructors should adopt a role of facilitator or resource rather than lecturer or grader.

The following learning pyramid illustrates the correlation between training method and learner retention:

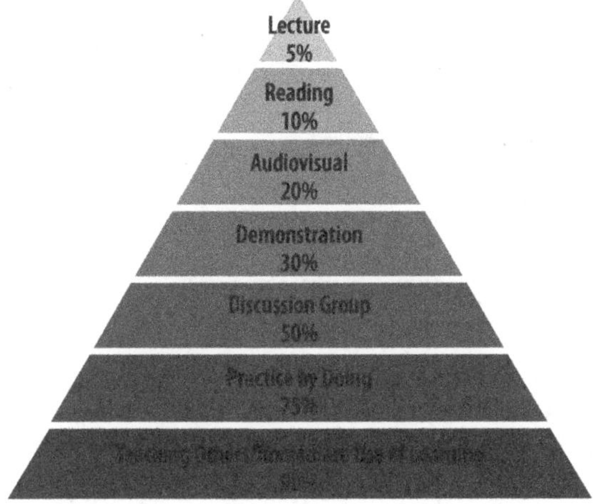

The Occupational Safety and Health Administration (OSHA) does not specifically address the responsibility of employers to provide health and safety information and instruction to employees, but Section 5(a)(2) of the OSH Act (General Duty Clause) does state that each employer "... shall comply with OSHA standards," and more than 100 of the standards contain training requirements. The adequacy of employee training may also become an issue in contested cases where the affirmative defense of unpreventable employee misconduct is raised. Under case law that is well-established in the Occupational Safety and Health Review Commission (OSHRC) and the courts, an employer may successfully defend against an otherwise valid citation by demonstrating that all feasible steps had been taken to avoid the occurrence of the hazard and that the actions of the employee involved in the violation were a departure from a uniformly and effectively enforced work rule of which the employee had either actual or constructive knowledge.

The OSHA Model

Determining Whether Training Is Needed
Identifying Training Needs
Identifying Goals and Objectives
Developing Learning Activities
Conducting the Training
Evaluating Program Effectiveness
Improving the Program

Learning objectives (LO) must be **measurable** and **observable.** They should have specific verbs (action words). They should describe something the trainer can see participants do or hear them say. They should also avoid using words that are difficult or impossible to measure. Learning objectives have three parts:

1. A behavior
2. A condition
3. A standard that is established early in the process

Measurable Learning Objectives

Appropriate Words (Observable/Measurable Behavior)		Inappropriate Words (Not Observable/ Nonmeasurable Behavior)	
write	explain	accept	appreciate
classify	list	be aware of	believe
calculate	select	remember	comprehend
prepare	apply	regard	know
operate	choose	be familiar with	understand
define	construct	consider	discern
describe	complete	grasp	ascertain
demonstrate		value	

Performance evaluations should be tied directly to the measurement of the skills, knowledge, and abilities defined in the learning objectives.

The documentation should include:

- Participant sign-in sheet/Training roster
- Course title and training outline
- Course date(s) and materials
- Statement or other proof that the student has successfully completed the training
- Name and address of the training provider (if appropriate)

It is essential to document all training for several reasons:
- Training records must be maintained to document that employees have received the proper training to perform their jobs safely.
- When inspectors visit a company, they usually review training records to determine whether training requirements have been met.
- When accurate records are not kept, training has to be repeated so that skills and knowledge can be documented.
- Proper documentation may be required before employees can work at certain sites or perform certain jobs.
- Training records keep trainers organized and assist with continuous improvement efforts.

Since OSHA requires trainers to be competent, it is a good idea to keep a record of trainer qualifications and credentials.

TRAINING METRICS

Researcher Donald Kirkpatrick's theory has now become arguably the most widely used and popular model for the evaluation of training and learning. Kirkpatrick's four-level model is now considered an industry standard across the HR and training communities. The four levels of Kirkpatrick's evaluation model essentially measure:
1. REACTIONS—What the students thought and felt about the training

2. LEARNING—The resulting increase in knowledge or capability
3. TRANSFER—The extent of behavior and capability improvement and implementation/application
4. RESULTS—The effects on the business or environment resulting from each trainee's performance

All four measures are recommended to gain full and meaningful evaluation of learning in organizations, but note that their application broadly increases in complexity, and usually cost, through the levels from level 1 to level 4.

Quantifying ROI means accounting for all the costs of the training program.

Fixed costs: Independent of the number of participants

Variable costs: Dependent on the number of participants

There are costs at every step—make sure to account for them all. Another useful and often used definition/formula expresses the ROI as the percentage return on the costs incurred. This has the advantage of communicating with many investors and stakeholders in their language. The formula to calculate ROI in this way is:

$$ROI(\%) = \frac{Benefit - Cost}{Cost} \times 100$$

Comparison of Common Evaluation Instruments

Instruments	Reaction	Learning	Behavior	Results	Advantages	Limitations
Questionnaire	✓		✓	✓	Low cost Honesty increased Anonymity optimized Respondent sets pace Variety of options	May not collect accurate information On-the-job responding conditions uncontrolled Respondents set pace No controllability of return rate
Attitude Survey	✓		✓	✓	Standardization possible Quickly processed Easily administered	Predetermined alternatives and response choices Reliance on norms may distort individual performance May not reflect true feelings
Written Test		✓			Low cost Readily scored Quickly processed Easily administered Wide sampling possible	May be threatening Possible low relation to job performance Reliance on norms may distort individual performance Possible cultural bias
Performance Test		✓	✓		Reliable Simulation potential Objective based	Time-consuming Simulation difficult High development costs

Comparison of Common Evaluation Instruments, continued

					Strengths	Weaknesses
Interview	✓		✓	✓	Flexible Opportunity for clarification Depth possible Personal contact	High reactive effects High costs Face-to-face threat potential Labor-intensive Trained interviewers needed
Focus Groups	✓		✓		Flexible Low cost Good qualitative responses Personal contact	Effectiveness depends on facilitator Subjective Sometimes difficult to summarize findings
Observation	✓		✓		Nonthreatening to participants Excellent way to measure behavior change	Possibly disruptive Reactive effect Unreliable Requires trained observers
Performance Records			✓	✓	Reliable Objective Job-based Ease of review Minimal reactive effects	Lack of knowledge of criteria for keeping/discarding records Information system discrepancies Indirect nature of data Need for conversion to usable forms Records prepared for other purposes Sometimes expensive to collect

Personal protective equipment (PPE) is the final line of defense for a worker to avoid exposure to a hazard. The hazard may potentially damage the respiratory tract, sight, hearing, hands, feet, or other parts of the body. Other hazard-abating techniques should be used before resorting to PPE. Engineering controls (such as filters, dampening devices, and guarding) and administrative controls (such as crew rotation and limited exposure time) should be attempted and enacted, if possible, before using PPE. If the hazard "breaks through" the PPE, nothing else can protect the worker.

The proper PPE is selected after a hazard analyses has been performed. Whenever a new process or new equipment is introduced into the work environment or whenever a current process is substantially changed, a hazard analysis should be performed to ensure that the worker will be protected.

When engineering controls and job hazard analyses do not eliminate all job hazards, employees will need (where appropriate) to wear PPE. This includes items such as caps, hairnets, face shields, safety goggles, glasses, hearing protection, foot guards, gloves, etc. Supervisors need to ensure that the equipment selected meets the following requirements:

- It is appropriate for the particular hazard.
- It is maintained in good condition.
- It is properly stored when not in use, to prevent damage or loss.
- It is kept clean, fully functional, and sanitary.

<table>
<tr><td colspan="3">Sample Supervisor's Safety Checklist—PPE

Supervisor: ___________________________________

Date: _____________

Department/Area: _______________________________</td></tr>
</table>

Yes	No	Checklist Area
		Employee Training—All workers have been trained to know
		When PPE is necessary.
		Which PPE is necessary.
		How to properly don, remove, adjust, and wear PPE.
		The limitations of the PPE.
		The proper care, maintenance, useful life, and disposal of the PPE.
		Area has been evaluated and PPE selected for hazards involving
		Eye and face injury
		Noise levels
		Cuts, punctures, and laceration
		Foot and hand injury
		Head impact
		Chemical exposure
		Air contamination
		Excessive temperatures

EYE AND FACE PROTECTION

In Appendix B of 29 *CFR* 1910, Subpart I, an eye and face protection chart is given to allow the employer or worker to select the correct face and eye protection based on the hazard encountered.

General industry standards of the Occupational Safety and Health Administration (OSHA) offer guidelines for eye protection while welding. The 29 *CFR* 1910.133 standard describes the minimum protective shades for arc welding based on electrode size and similar guidelines based on plate thickness.

Eye and Face Protection Selection Chart

Source	Assessment of Hazard	Protection
Impact—Chipping, grinding machining, masonry work woodworking, sawing, drilling, chiseling, powered fastening, riveting, and sanding.	Flying fragments, objects, large chips, particles, sand, dirt, etc.	Spectacles with side protection, goggles, face shields. See notes (1), (3), (5), (6), (10). For severe exposure, use face shield.
Heat—Furnace operations, pouring, casting, hot dipping, and welding.	Hot sparks	Face shields, goggles, spectacles with side protection. For severe exposure use face shield. See notes (1), (2), (3).
	Splash from molten metals	Face shields worn over goggles. See notes (1), (2), (3).
	High-temperature exposure	Screen face shields, reflective face shields. See notes (1), (2), (3).
Chemicals—Acid and chemicals handling, degreasing plating.	Splash	Goggles, eyecup and cover types. For severe exposure, use face shield. See notes (3), (11).
	Irritating mists	Special-purpose goggles.
Dust—Woodworking, buffing, general dusty conditions.	Nuisance dust	Goggles, eyecup and cover types. See note (8).
Light and/or Radiation—		
Welding: Electric arc	Optical radiation	Welding helmets or welding shields. Typical shades: 10–14. See notes (9), (12).
Welding: Gas	Optical radiation	Welding goggles or welding face shield. Typical shades: gas welding 4–8, cutting 3–6, brazing 3–4, See note (9).
Cutting, Torch brazing, Torch soldering	Optical radiation	Spectacles or welding face shield. Typical shades, 1.5–3. See notes (3), (9).
Glare	Poor vision	Spectacles with shaded or special purpose lenses, as suitable. See notes (9), (10).

Notes to Eye and Face Protection Selection Chart

1. Care should be taken to recognize the possibility of multiple and simultaneous exposure to a variety of hazards. Adequate protection against the highest level of each of the hazards should be provided. Protective devices do not provide unlimited protection.
2. Operations involving heat may also involve light radiation. As required by the standard, protection from both hazards must be provided.
3. Face shields should be worn only over primary eye protection (spectacles or goggles).
4. As required by the standard, filter lenses must meet the requirements for shade designations in §1910.133(a)(5). Tinted or shaded lenses are not filter lenses unless they are marked or identified as such.
5. As required by the standard, persons whose vision requires the use of prescription (Rx) lenses must wear either protective devices fitted with prescription (Rx) lenses or protective devices designated to be worn over regular prescription (Rx) eyewear.
6. Wearers of contact lenses must also wear appropriate eye and face protection in a hazardous environment. It should be recognized that dusty and/or chemical environments may represent an additional hazard to contact lens wearers.
7. Caution should be exercised in the use of metal frame protective devices in electrical hazard areas.
8. Atmospheric conditions and the restricted ventilation of the protector can cause lenses to fog. Frequent cleansing may be necessary.
9. Welding helmets or face shields should be used only over primary eye protection (spectacles or goggles).
10. Non-sideshield spectacles are available for frontal protection only and are not acceptable eye protection for the source and operations listed for "impact."

11. Ventilation should be adequate but well protected from splash entry. Eye and face protection should be designed and used so that it both provides adequate ventilation and protects the wearer from splash entry.
12. Protection from light radiation is directly related to filter lens density. See note (4). Select the darkest shade that allows task performance.

Filter Lenses for Protection against Radiant Energy

Operations	Electrode Size ⅟₃₂ in.	Arc Current	Minimum* Protective Shade
Shielded metal arc welding	Less than 3 3–5 5–8 More than 8	Less than 60 60–160 160–250 250–550	7 8 10 11
Gas metal arc welding and flux cored arc welding		Less than 60 60–160 160–250 250–500	7 10 10 10
Gas tungsten arc welding		Less than 50 50–150 150–500	8 8 10
Air carbon	(Light)	Less than 500	10
Arc cutting	(Heavy)	500–1000	11
Plasma arc welding		Less than 20 20–100 100–400 400–800	6 8 10 11
Plasma arc cutting	(Light)** (Medium)** (Heavy)**	Less than 300 300–400 400–800	8 9 10
Torch brazing			3
Torch soldering			2
Carbon arc welding			14

Filter Lenses for Protection against Radiant Energy

Operations	Plate thickness (inches)	Plate thickness (inches)	Minimum* Protective Shade
Gas welding:			
Light	Under ⅛	Under 3.2	4
Medium	⅛ to ½	3.2 to 12.7	5
Heavy	Over ½	Over 12.7	6
Oxygen cutting:			
Light	Under 1	Under 25	3
Medium	1 to 6	25 to 150	5
Heavy	Over 6	Over 150	3

* As a rule of thumb, start with a shade that is too dark to see the weld zone. Then go to the lighter shade that gives sufficient view of the weld zone without going below the minimum. In oxyfuel gas welding or cutting, where the torch produces a high yellow light, it is desirable to use a filter lens that absorbs the yellow or sodium line in the visible light of the (spectrum) operation.

** These values apply where the actual arc is clearly seen. Experience has shown that lighter filters may be used when the arc is hidden by the workpiece.

The construction standard, 29 *CFR* 1926.103, offers a small chart to select eye protection based on the power of the laser that a worker will be using.

HEAD PROTECTION

Head protection guidelines are provided by the American National Standards Institute (ANSI) and incorporated into the OSHA personal protective equipment requirements. A hard hat must be able to protect the worker from an overhead hazard. The standard is the protection equivalent of an 8-lb ball dropped 5 feet. Hard hats have a bill in the front to protect the worker's nose and face. The hard hat must be worn with the bill facing forward and the internal harness oriented the correct way. Welder's hard hats are typically the only head protection that may be worn in either direction—with the unbilled side to be used under the welding shield.

Types of Protective Headwear

	Voltage Protection	Comments
Type G	Low voltage protection—2,200 volts	General
Type E	High voltage protection—22,000 volts	Electrical
Type C	No voltage protection, metal	Conductive
Type I	Intended to protect from blows to the top of the head	
Type II	Intended to protect from blows to the top of the head and side impact	

Reference: After ANSI Z89.1

LUNG FUNCTION AND RESPIRATION

External respiration is the exchange of oxygen (O_2) and (CO_2) between blood and the air in the lungs. Internal respiration is the exchange of O_2 and CO_2 between the blood surrounding the tissue and the tissue cells. The lungs' defense mechanisms are phagocytes (which envelop the invading particle), the mucous lining (which prevents the particle from infiltrating very deeply), and muscular constriction (which tries to expel the particle). Particle sizes that are retained in the lungs vary from 1 to 5μ, with those particles < 1μ traveling into the alveoli. Respirable dust is that part of the total dust amount that can enter into the lungs. Respirable dust is generally the smaller-sized components of the total dust amount. At rest, the average adult inhales approximately 5 liters of air every minute (500 cc per breath × 10 breaths per minute = 5,000 cc per minute = 5 liters per minute).

RESPIRATORS

Respirators either filter out contaminants in the ambient air or provide contaminant-free air that is supplied through outside means. The most common type of respirator is the air-purifying respirator (APR), which uses filters to collect and trap specific contaminants and then convey the filtered air to the worker. APRs may be powered either by the force of the worker's lungs (negative pressure) or by the use of a battery and motor arrangement (positive pressure). They are generally inexpensive, reliable if used correctly, and relatively manageable. However, they are able to convey only the same quality of air that is in the breathing area. If this air is not capable of sustaining life, then APRs cannot increase the percentage of oxygen but only filter out the specific contaminant. Oxygen-deficient air is generally considered to be air with less than 19.5% oxygen, although the American Conference of Governmental Industrial Hygienists (ACGIH) level is less than 18%.

Another type of respirator is the supplied-air respirator (SAR). These respirators supply air to the worker via tanks worn on the worker's back, via tank systems and hose, or via compressor and hose.

If the worker wears the tanks, this is referred to as a self-contained breathing apparatus (SCBA). Similarly, the tanks that underwater divers use are called SCUBA for "self-contained underwater breathing apparatus." An open-loop SCBA exhausts used air, while a closed-loop SCBA recycles the exhaled air after CO_2 and H_2O have been removed.

Both the air in the tanks and the air in the hose systems have to meet certain quality criteria. If an air compressor is used, it must be oil-less to prevent fumes from entering the air system. Grade D air is supplied, with a maximum of 20 ppm CO and a maximum of 1,000 ppm CO_2.

Some atmospheres cannot support life in any manner. These atmospheres are called immediately dangerous to life and health (IDLH). In such atmospheres, a self-contained positive-pressure respirator is to be used.

Respirator Filter Cartridge Colors

Contaminant	Color Coding on Cartridge/Canister
Acid gases	White
Hydrocyanic acid gas	White with ½-inch green stripe completely around the canister near the bottom
Chlorine gas	White with ½-inch yellow stripe completely around the canister near the bottom
Organic vapors	Black
Ammonia gas	Green
Acid gases and ammonia gas	Green with ½-inch white stripe completely around the canister near the bottom
Carbon monoxide	Blue
Acid gases and organic vapors	Yellow
Hydrochloric acid gas and chloropicrin vapor	Yellow with ½-inch blue stripe completely around the canister near the bottom
Acid gases, organic vapors, and ammonia gases	Brown
Radioactive materials, except tritium and noble gases	Purple (magenta)
Pesticides	Organic vapor canister plus a particulate filter
Multi-Contaminant and CBRN agent	Olive
Any particulates—P-100	Purple
Any particulates—P-95, P-99, R-95, R-99, R-100	Orange
Any particulates free of oil—N-95, N-99, or N-100	Teal

FOOT PROTECTION

Employees exposed to the hazards of rolling or falling objects, sharp objects that can penetrate the sole of footwear, static electricity buildup, or contact with energized electrical conductors require protective footwear (29 *CFR* 1910.136). Specific guidelines are referenced in ANSI Z41, as follows:

- Toe Guard Footwear (impact and compression requirements)
- Metatarsal Guard Footwear
- Conductive Footwear
- Electrical Hazard Footwear
- Sole Puncture Resistant Footwear
- Electrostatic Dissipative Footwear

Modified Chart of EPA/OSHA Levels of Protection

Levels	Skin	Respiratory	When
A	Fully encapsulating, chemical-resistant suit, inner gloves, chemical-resistant safety boots	Pressure-demand, full-facepiece SCBA or pressure-demand supplied-air respirator with escape SCBA	Highest level of protection indicated by high concentration of atmospheric vapors, gases, or particulates or splash hazard exists.
B	Chemical-resistant clothing (overalls and long-sleeved jacket; hooded, one- or two-piece chemical splash suit; disposable, chemical-resistant one-piece suit), inner and outer gloves, chemical-resistant safety boots and hard hat	Pressure-demand, full-facepiece SCBA or pressure-demand supplied-air respirator with escape SCBA	High level of respiratory protection required, but less skin protection. IDLH, less than 19.5% oxygen.
C	Chemical-resistant clothing (overalls and long-sleeved jacket; hooded, one- or two-piece chemical splash suit; disposable, chemical-resistant one-piece suit), inner and outer gloves, chemical-resistant safety boots and hard hat	Full-facepiece, air-purifying, canister-equipped respirator	The contaminants, splashes, or direct contact will not affect exposed flesh. Canister will remove contaminant.
D	Overalls, safety boots, safety glasses or chemical splash goggles, hard hat	No respiratory protection and minimal skin protection	The atmosphere contains no known hazard. Splashes, immersion, or inhalation improbable.

Assigned Protection Factors

Type of Respirator[1,2]	Quarter mask	Half mask	Full facepiece	Helmet/ Hood	Loose-fitting facepiece
1. Air-Purifying Respirator	5	10^3	50	—	—
2. Powered Air-Purifying Respirator (PAPR)	—	50	1,000	25/1,000[4]	25
3. Supplied-Air Respirator (SAR) or Airline Respirator					
• Demand mode	—	10	50	—	—
• Continuous flow mode	—	50	1,000	25/1,000[4]	25
• Pressure-demand or other positive-pressure mode	—	50	1,000	—	—
4. Self-Contained Breathing Apparatus (SCBA)					
• Demand mode	—	10	50	50	—
• Pressure-demand or other positive-pressure mode (e.g., open/closed circuit)	—	—	10,000	10,000	—

Notes:

[1] Employers may select respirators assigned for use in higher workplace concentrations of a hazardous substance for use at lower concentrations of that substance, or when required respirator use is independent of concentration.

[2] The assigned protection factors in *Table I* are only effective when the employer implements a continuing, effective respirator program as required by this section (29 *CFR* 1910.134), including training, fit testing, maintenance, and use requirements.

[3] This APF category includes filtering facepieces, and half masks with elastomeric facepieces.

[4] The employer must have evidence provided by the respirator manufacturer that testing of these respirators demonstrates performance at a level of protection of 1,000 or greater to receive an APF of 1,000. This level of performance can best be demonstrated by performing a WPF or SWPF study or equivalent testing. Absent such testing, all other PAPRs and SARs with helmets/hoods are to be treated as loose-fitting facepiece respirators, and receive an APF of 25.

[5] These APFs do not apply to respirators used solely for escape. For escape respirators used in association with specific substances covered by 29 *CFR* 1910 subpart Z, employers must refer to the appropriate substance-specific standards in that subpart. Escape respirators for other IDLH atmospheres are specified by 29 *CFR* 1910.134(d)(2)(ii).

Example Personal Protective Equipment Audit

Facility: _______________________ Area: _______________________

Auditor: _______________________ Date: _______________________

Area	Satisfactory	Action Required	Corrective Action (date)
Employee Knowledge			
Date of last PPE training			
When to use PPE			
Limitations			
Selection and inspection			
Cleaning and storage			
Donning and removal			
Program Administration			
Hazard assessment completed			
Hazard control survey completed			
PPE hazard certification completed			
High hazard areas identified			
PPE disposal procedures discussed			
Safeguards			
Engineering safeguards			
Administrative safeguards			
Training safeguards			

Example Personal Protective Equipment Audit, continued

Area Inspection			
Signs and warnings posted			
Adequate PPE stock available			
Electricians wear electrically rated safety shoes/hard hats			
PPE clean and properly stored			
PPE used properly			

Operational Questions	
	Is a hazard assessment procedure used to determine whether hazards that require the use of personal protective equipment (for example, head, eye, face, hand, or foot protection) are present or are likely to be present?
	If hazards or the likelihood of hazards is found, is PPE selected and are affected employees properly fitted for the personal protective equipment suitable for protection from these hazards?
	Have employees been trained on PPE procedures; that is, what PPE is necessary for a job task, when they need it, and how to properly adjust it?
	Are protective goggles or face shields provided and worn where there is danger of flying particles or corrosive materials?
	Are approved safety glasses required to be worn at all times in areas where there is a risk of eye injuries such as punctures, abrasions, contusions, or burns?
	Are employees who need corrective lenses (glasses or contacts) and who work in environments having harmful exposures required to wear only approved safety glasses or protective goggles or to use other medically approved, precautionary procedures?

Example Personal Protective Equipment Audit, continued

		Are protective gloves, aprons, shields, or other means provided and required where employees could be cut or where there is reasonably anticipated exposure to corrosive liquids, chemicals, blood, or other potentially infectious materials?
		Are hard hats provided and worn where the danger of falling objects exists?
		Are hard hats inspected periodically for damage to the shell and suspension system?
		Is appropriate foot protection required where there is a risk of foot injuries from hot, corrosive, or poisonous substances, falling objects, or crushing or penetrating actions?
		Are approved respirators provided for regular or emergency use where needed?
		Is all personal protective equipment maintained in a sanitary condition and ready for use?
		Are there eyewash facilities and a quick drench shower within the work area where employees are exposed to injurious, corrosive materials? Where special equipment is needed for electricians, is it available?
		Where food or beverages are consumed on the premises, are they consumed in areas where there is no exposure to toxic material, blood, or other potentially infectious materials?
		Is protection against the effects of occupational noise exposure provided when sound levels exceed those of the OSHA noise standard?
		Are adequate work procedures, protective clothing, and equipment provided and used when cleaning up spilled toxic or otherwise hazardous materials or liquids?
		Are appropriate procedures in place for disposing of or decontaminating personal protective equipment contaminated with, or reasonably anticipated to be contaminated with, blood or other potentially infectious materials?
NOTES:		

Audits and inspections usually involve some criterion-referenced standards. They can be based on regulatory compliance or internal conformance with policies and procedures. They can be conducted both internally and externally. Scheduling an audit may depend on:

- Whether mandated or not
- Frequency of injuries
- Probability of injuries
- Potential loss severity
- Potential for damage
- New/modified equipment, processes, operations

Procedures similar to an audit are assessments, evaluations, and analyses. Although these approaches may consider compliance or conformance, auditing revolves around subjective and objective criteria, not just a specific performance standard.

During the site visit there will be a finite amount of time to achieve the objectives of the audit plan. This makes it important to be knowledgeable about the organization, its operations, and, in the case of a compliance audit, the facility's regulatory status before arriving on-site so that no time is lost. Get a thorough description of the facility from facility personnel in order to prepare. Knowledge of environmental regulations is crucial to conducting any type of environmental audit. During a management system audit, the focus is on the system and its policies and procedures so that a decision can be made whether or not the requirements of the standard and the intent of management have been implemented. An identified compliance deficiency can be important evidence of a system's problem. In a compliance audit, you must know the details of the regulations because a compliance deficiency is what you are searching for. Also for a compliance audit, conduct a preliminary review of state and local regulations in order to determine the scope of regulations that may apply to the facility. Many states regulate local environmental issues that are not addressed at the federal level. Be aware also that the state and local regulations may vary significantly in detail. In a compliance audit, an on-site review of documents may be necessary since there will probably be too many records to inspect prior to the site visit in order to verify that the operations are in compliance with regulatory requirements. The more information that is reviewed before the visit, the more effective and efficient the audit team can be during the visit. A previsit document review will be a time-saving benefit and will minimize the disruption of operations that usually occurs during an audit. Steps for planning include:

1. Decide what to inspect.
2. Create a safety inspection checklist.
3. Determine how to conduct the inspections.
4. Determine when to inspect.
5. Determine who should inspect.
6. Train the inspectors.

Other considerations include:
- Use interpersonal skills to work with managers and employees.
- Provide clarification to employees about safety inspections.
- Gather information, including how to identify the hazards.
- Take notes and record the information you are gathering.

LEADING AND LAGGING INDICATORS

Lagging indicators measure a company's incidents in the form of past accident statistics. Lagging indicators are the traditional safety metrics used to indicate progress toward compliance with safety rules. These are the bottom-line numbers that evaluate the overall effectiveness of safety at your facility. They tell you how many people got hurt and how badly.

Examples include:
- Injury frequency and severity
- OSHA recordable injuries
- Lost workdays
- Worker's compensation costs

A leading indicator is a measure preceding or indicating a future event used to drive and measure activities carried out to prevent and control injury. Leading indicators are focused on future safety performance and continuous improvement. These measures are proactive in nature and report what employees are doing on a regular basis to prevent injuries.

Examples include:
- Safety training
- Hazards identified and corrected
- Safe Work Practice Observations
- Employee perception surveys
- Safety conformance or compliance audits

Document Request for Auditing

Compliance and Management System Auditing	Management System Auditing
Organization chart identifying the names and titles of key personnel	Vision/mission statements and strategic plan
Identification of departments and operating groups, with a description of the operations conducted	Environmental manual with ISO 14001 policies and management-level procedures that address the core elements of the EMS
General site-specific information describing number of employees, square feet of building space, location	Current EMS standard operating procedures (SOP) or work instructions for environmental requirements
Site plot plan showing buildings and setup of processes inside the buildings	Sustainability program and list of long-term sustainability objectives
Environmental policy	List of environmental aspects and impacts
Comprehensive list of buildings	Environmental performance metrics and identification of the environmental performance data that is currently collected
Process descriptions and flowcharts for those with the greatest environmental risk	List of set environmental objectives and targets and the environmental management programs developed to achieve environmental objectives
Lists of raw materials, products, by-products, and waste streams	
List of environmental permits	
List of environmental plans and best management practices	
Training matrix showing environmental training required for job positions	
Most recent internal and/or external environmental audit reports	

Sample Periodic Internal Safety Audit

For items checked "No," fill out a Maintenance Work Order. Mark "N/A" for items not applicable to your area.

Fire Protection	Yes	No
Fire extinguishers inspected, charged, accessible (3-ft clearance)		
Combustible material removed and stored properly; flammable material in approved areas		
Exit routes clear and EXIT or NO EXIT signs posted (lighted and visible)		
Evacuation routes posted		
Storage separated from walls and ceiling (18" minimum for sprinklered areas)		
Electrical Safety		
Power panels, controls, receptacles, and wiring covered; no missing, loose, or broken parts		
Electric power cords not frayed or broken; all plugs with three prongs		
No extension cords through walls, doors, ceiling, or windows or under mats or rugs		
Electric panels marked to indicate Service and Voltage; 3-foot clearance on each side		
Trip-Slip-Fall Hazards		
Drain covers and grates in good repair and installed		
Walkways clear of material and cords		
Guardrails and steps secured; ladders in good repair, no missing or loose parts		
Adequate lighting in all areas, including exterior night lighting		
Personal Protection		
Machine guards in place		
Emergency eyewash stations capped, functional, accessible		
Personal protective equipment being used		
Good body mechanics (lifting, pushing, pulling, range of motion, no twisting)		
Lockout/tagout program properly used		

Chemical Safety		
All containers properly labeled with specific hazards and closed/sealed		
Only the minimum amount needed in the work area; all other chemicals properly stored		
Forklifts and Pallet Jacks		
All operators with current licenses		
All equipment functioning properly—brakes, horn, controls, backup alarm		
Forklift seat belts used		
Traffic routes established and marked		

Route to: ____

Department Superintendent: _______________________________

Safety Office: ____

OSHA INSPECTIONS, CITATIONS, AND PENALTIES

The Occupational Safety and Health Administration (OSHA) was created as part of the Occupational Safety and Health Act (OSH Act) of 1970. It is part of the U.S. Department of Labor. Individual states may have similar agencies that safeguard worker health and safety. Depending on the state, the state agency may or may not supersede the federal OSHA.

One of the most important sections of the OSH Act is Section 5, which outlines the duties of both employers and employees. Section 5(a)(1) is commonly referred to as the General Duty Clause and is addressed to employers, while Section 5(b) refers to the obligations of employees.

5. Duties

(a) Each employer (1) shall furnish to each of his employees employment and a place of employment which are free from recognized hazards that are causing or are likely to cause death or serious physical harm to his employees; (2) shall comply with occupational safety and health standards promulgated under this Act. (b) Each employee shall comply with occupational safety and health standards and all rules, regulations, and orders issued pursuant to this Act which are applicable to his own actions and conduct.

Reference: OSH Act of 1970

OSHA proposes standards that establish the minimum acceptable level of protection for the topic area. Vertical standards are specific to one industry while horizontal standards cover many industries. A specification standard details specifications for compliance while a performance standard leaves the details of compliance to the industry.

The OSH Act and OSHA have a compliance aspect. Compliance officers visit companies and worksites to monitor compliance with the OSHA standards. The visit may be part of a target initiative, a worker complaint, or a random inspection.

OSHA seeks to focus its inspection resources on the most hazardous workplaces in the following order of priority:

1. **Imminent danger situations**—Hazards that could cause death or serious physical harm receive top priority. Compliance officers will ask employers to correct these hazards immediately or remove endangered employees.
2. **Fatalities and catastrophes**—Incidents that involve a death or the hospitalization of three or more employees come next. Employers must report such catastrophes to OSHA within 8 hours of their occurrence.
3. **Complaints**—Allegations of hazards or violations also receive a high priority. Employees may request anonymity when they file complaints.
4. **Referrals** of hazard information from other federal, state, or local agencies, individuals, organizations, or the media receive consideration for inspection.

5. **Follow-ups**—Checks for abatement of violations cited during previous inspections are also conducted by the agency in certain circumstances.
6. **Planned or programmed investigations**—Inspections aimed at specific, high-hazard industries or individual workplaces that have experienced high rates of injuries and illnesses also receive priority.

Phone/Fax Investigations

OSHA prioritizes all complaints it receives based on their severity. For lower-priority hazards, with permission of a complainant, OSHA may telephone the employer to describe safety and health concerns, following up with a fax providing details on alleged safety and health hazards. The employer must respond in writing within five working days, identifying any problems found and noting corrective actions taken or planned. If the response is adequate and the complainant is satisfied with the response, OSHA generally will not conduct an on-site inspection.

ON-SITE INSPECTIONS

- **Preparation**—Before conducting an inspection, OSHA compliance officers research the inspection history of a worksite using various data sources and review the operations and processes in use and the standards most likely to apply. They gather appropriate personal protective equipment and testing instruments to measure potential hazards.
- **Presentation of credentials**—The on-site inspection begins with the presentation of the compliance officer's credentials, which include both a photograph and a serial number.
- **Opening Conference**—The compliance officer will explain why OSHA selected the workplace for inspection and describe the scope of the inspection, walkaround procedures, employee representation, and employee interviews. The employer then selects a representative to accompany the compliance officer

during the inspection. An authorized representative of the employees, if any, also has the right to go along. In any case, the compliance officer will consult privately with a reasonable number of employees during the inspection.

- **Walkaround**—Following the opening conference, the compliance officer and the representative(s) will walk through the portions of the workplace covered by the inspection, looking for hazards that could lead to employee injury or illness. The compliance officer will also review worksite injury and illness records and the posting of the official OSHA poster. During the walkaround, compliance officers may point out some apparent violations that can be corrected immediately. Although the law requires that these hazards still be cited, prompt correction is a sign of good faith on the part of the employer. Compliance officers try to minimize work interruptions during the inspection and will keep confidential any trade secrets they observe.
- **Closing Conference**—After the walkaround, the compliance officer holds a closing conference with the employer and the employee representative(s) to discuss the findings. The compliance officer discusses possible courses of action that the employer may take following the inspection, which could include having an informal conference with OSHA or contesting citations and proposed penalties.

Results

OSHA must issue a citation and proposed penalty within six months of the violation's occurrence. Citations describe the OSHA requirements allegedly violated, list any proposed penalties, and give a deadline for correcting the alleged hazards. Violations are categorized as other-than-serious, serious, willful, repeated, and failure to abate. Penalties may range up to $7,000 for each serious violation and up to $70,000 for each willful or repeated violation. Penalties may be reduced based on an employer's good faith, inspection history, and size of business or may be increased for "bad actors." For serious violations, OSHA may also increase the proposed penalty based on the gravity of the alleged violation.

Appeals

When OSHA issues a citation to an employer, it also offers the employer an opportunity for an informal conference with the OSHA area director to discuss citations, penalties, abatement dates, or any other information pertinent to the inspection. The agency and the employer may work out a settlement agreement to resolve the matter and to eliminate the hazard. OSHA's primary goal is the enforcement of federal standards. Employers have 15 working days after receipt of citations and proposed penalties to formally contest the alleged violations and/or penalties by sending a written notice to the area director. OSHA then forwards the contestation to the Occupational Safety and Health Review Commission (OSHRC) for independent review. Alternatively, citations, penalties, and abatement dates that are not challenged by the employer or are settled become a final order of the Occupational Safety and Health Review Commission.

Record Keeping

Occupational injuries and illnesses must be tracked by the employer on the OSHA 301 form. Each year's total record must be posted during the month of February of the following year on the OSHA log, known as the OSHA 300 form. The log is then retained for five additional years. The requirement to keep occupational injury and illness records applies to all companies except those considered small establishments (no more than 10 employees in the entire company) or those in low-hazard industries (by SIC Code). Employers must post an occupational injury or illness within six days of the knowledge of the injury/illness. The injury/illness must be categorized correctly and any lost or restricted time catalogued. In the case of a fatality, the local OSHA office must be immediately notified. If an incident results in five hospitalizations, notification is also required. Refer to OSHA publication 3245-09R 2005 for more information.

Guide to Determining Whether to Record an Illness or Injury

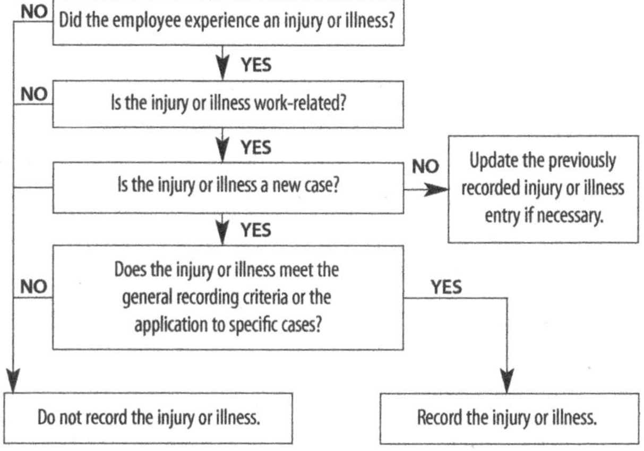

Accident prevention and control of hazards are the results of a well-designed and executed safety and health program. The keys to a successful program include unbiased, prompt, and accurate incident investigations. The basic purpose of these investigations is to determine measures that can be taken to prevent the occurrence of similar incidents in the future. This chapter addresses:

- Company policy
- Responsibilities
- Hazard control
- Role of supervisors
- Investigation procedures

Investigations of all work-related incidents, injuries, and illnesses are to be conducted in a professional manner in order to identify probable causes. Investigations are also used to develop specific management actions to prevent future incidents.

RESPONSIBILITIES

Immediate Steps:
1. Arrive safely.
2. Provide first aid for any injured persons.
3. Eliminate or control hazards.
4. Supervise the scene and preserve evidence.
5. Document accident scene information to determine the cause.
6. Interview witnesses immediately.

Management should:
- Conduct accident prevention and investigation training for supervisors.
- Ensure that all accidents and injuries are properly investigated.

- Ensure that immediate and long-term corrective actions are taken to prevent recurrence.
- Permanently maintain accident reports on file.
- Ensure that proper entries are made in the OSHA 300 Log at the first report of injury.
- Provide all necessary medical care for injured workers.
- Lead investigations of any fatalities or catastrophic incidents.

Supervisors should:
- Conduct immediate initial incident investigations.
- Report all accidents to management as soon after they occur as possible.
- Collect and preserve all evidence that may be useful in an investigation.
- Conduct interviews of witnesses in a polite, professional manner.
- Refrain from attempting to find or assign blame for accidents.
- Take action to protect people and property from secondary effects of accidents.

Employees should:
- Immediately report all incidents and injuries to their supervisors.
- Assist as requested in all incident investigations.
- Report all hazardous conditions and near misses to supervisors.

INCIDENT PREVENTION

Incidents are usually complex. An incident may have 10 or more events that can be causes. A detailed analysis of an accident will normally reveal three cause levels: basic, indirect, and direct. At the lowest level, an accident results only when a person or an object receives an amount of energy or hazardous material that cannot be absorbed safely. This energy or hazardous material is the *direct cause* of the incident. The direct cause is usually the result of one or more unsafe acts or unsafe conditions, or both. Unsafe acts and conditions are the *indirect causes* or symptoms. In turn, indirect causes are usually traceable to poor management policies and decisions or to personal or environmental factors. These are the *basic causes*.

Most incidents are preventable by eliminating one or more causes. For this reason, it is important for incident investigations to determine not only what happened but also how and why. The information gained from these investigations can prevent recurrence of similar or perhaps more disastrous accidents. Incident investigators are interested in each event as well as the sequence of events that led to the accident. The accident type is also important to the investigator. The recurrence of incidents of a particular type or those with common causes identifies areas that need special incident prevention focus.

INITIAL INVESTIGATION PROCEDURES

The initial investigation has three purposes:
1. Prevent further possible injury and property damage
2. Collect facts about the accident
3. Collect and preserve evidence

Steps
a. Secure the area. Do not disturb the scene unless a hazard exists.
b. Prepare the necessary sketches and photographs. Label each item carefully and keep accurate records.
c. Interview each victim and witness. Also interview those who were present before the accident and those who arrived at the site shortly after the accident. Keep accurate records of each interview. Use a tape recorder if desired and if approved.

Determine the following:
a. What was abnormal before the accident.
b. Where the abnormality occurred.

    c. When it was first noted.
    d. How it occurred.

FOLLOW-UP ACCIDENT INVESTIGATION

The follow-up investigation is used to analyze data and determine the causes and corrective actions necessary to prevent recurrence.

Steps
    a. Analyze the data obtained in the initial investigation.
    b. Repeat any of the prior steps if necessary.
    c. Determine the following:
        1. Why the accident occurred.
        2. The likely sequence of events and probable causes (direct, indirect, basic).
    d. Identify the most likely causes.
    e. Conduct a post-investigation briefing.
    f. Prepare a summary report, including the recommended actions to prevent a recurrence.

An investigation is not complete until all data is analyzed and a final report is issued. In practice, the investigative work, data analysis, and report preparation proceed simultaneously during much of the time spent on the investigation.

CONDUCTING INTERVIEWS

In general, experienced personnel should conduct interviews. All interviews should be conducted in a quiet and private location. It is essential to get preliminary statements as soon as possible from all witnesses. Interviewers should not provide any facts to the witness but ask only nonleading or open-ended questions. In addition, the interviewer should:
    1. Explain the purpose of the investigation (accident prevention) and put each witness at ease.
    2. Listen, let each witness speak freely, and be professional, courteous, and considerate.

3. Take notes without distracting the witness and use a tape recorder only with consent of the witness.
4. Use sketches and diagrams to help the witness recall the scene or describe where he or she saw certain details.
5. Emphasize the areas of direct observation. Label hearsay accordingly.
6. Refrain from arguing with the witness.
7. Record the exact words the witness uses to describe each observation.
8. Identify each witness (name, address, occupation, years of experience, etc.).

Uses of Questions during Feedback

Uses	Samples
Get Discussion Started	What are the main purposes of . . . ? How would you define . . . ? What is your opinion of . . . ?
Capture and Maintain Interest	Safety! What does the term mean to you? How can you evaluate safety on the job? Which of these seems most useful? Why?
Keep Group Give-and-Take Going to Generate Ideas and Information	Interesting point! How do you feel about this? Are there additional points to consider? Now that we've considered the positive, what are some negative points of this idea?
Get Nonparticipants Involved	What's your opinion, Beth? Where else have you seen this occur, Tom? How does this affect your workplace, Derek?
Clarify Information or Comments	Can you give us an example of this point, Jim? How would you paraphrase Jim's point, Maria?
Keep Discussion on Target	Interesting point. Can you relate this to an earlier point? How does this relate to the original question?
Lead Discussion to Next Segment	Now, how does safety relate to job performance?

Assess and Evaluate Information	Given this data, what conclusions can we draw?
Suggest a Desired Response	Might there be a better way? What's the first, most obvious thing that should be done?
Assess Understanding/Get Feedback	What is the first step in this process? Why is it necessary to follow the steps in sequence? What problems can arise if you skip Step 2?
Get Agreement or Conclusion	What conclusion can we draw from this? What is the end result of all this? How will all this contribute to your safety?

INCIDENT ANALYSIS

Accidents represent problems that must be solved through investigations. Formal procedures are helpful in identifying and solving problems. This section discusses two of the most common procedures: change analysis and job safety analysis.

CHANGE ANALYSIS

As its name implies, this technique emphasizes change. To solve a problem, an investigator must look for deviations from the norm; assume that all the effects result from some unanticipated change; and analyze the change to determine its causes. Use the following steps in this procedure:

1. Define the problem. (What happened?)
2. Establish the norm. (What should have happened?)
3. Identify, locate, and describe the change. (What, where, when, to what extent?)
4. Specify what was and what was not affected.
5. Identify the distinctive features of the change.
6. List the possible causes.
7. Select the most likely causes.

JOB SAFETY ANALYSIS

Job safety analysis (JSA), which determines the events and conditions that led to an accident, is part of many existing accident prevention programs. In general, JSA breaks down a job into basic steps, identifies the hazards associated with each step, and prescribes controls for each hazard. A JSA also includes a chart listing these steps, hazards, and controls. Review the JSA during the investigation if a JSA has been conducted and a chart compiled for the job involved in the accident. Compile a JSA if one is not available.

INVESTIGATION REPORT

An accident investigation is not complete until a report has been prepared and submitted to management. To be an effective tool, an accident report should be clear and concise. The following outline has been found especially useful in compiling the information to be included in the formal report:

1. Background information
   a. Where and when the accident occurred
   b. Who and what were involved
   c. Operating personnel and other witnesses
2. Account of the accident (What happened?)
   a. Sequence of events
   b. Extent of damage
   c. Accident type
   d. Agency or source (of energy or hazardous material)
3. Discussion (analysis of the accident: who, what, when, where, and the five whys)
   a. Direct causes (energy sources, hazardous materials)
   b. Indirect causes (unsafe acts and conditions)
   c. Basic causes (management policies, personal or environmental factors)

4. Recommendations for immediate and long-range actions to remedy:
   a. Basic causes
   b. Indirect causes
   c. Direct causes (such as reduced quantities of protective equipment or poorly conceived workstation design)

POSSIBLE CAUSES

Obvious accident causes are most probably symptoms of a "root cause" problem. Some examples of unsafe acts and unsafe conditions that may lead to accidents are:

Unsafe Acts
- Unauthorized operation of equipment
- Running; horseplay; not following procedures; by-passing safety devices
- Not using protective equipment
- Being under the influence of drugs or alcohol

Unsafe Conditions
- Ergonomic hazards
- Environmental hazards
- Inadequate housekeeping
- Blocked walkways
- Improper or damaged PPE
- Inadequate machine guarding

Fishbone Diagram

Purpose: Problem's Root Causes

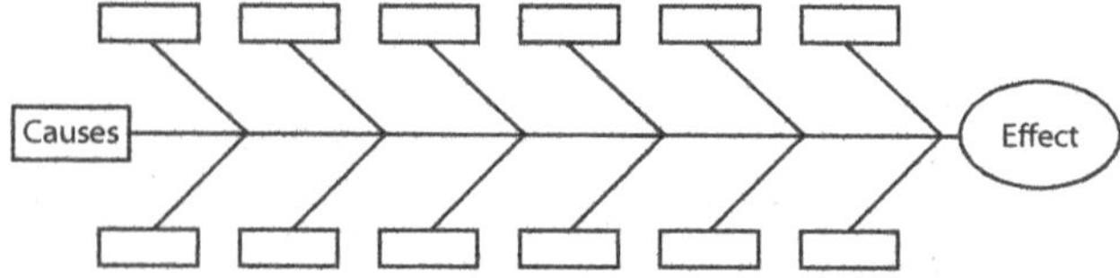

RECOMMENDATIONS

The primary purpose of incident investigation is to prevent the recurrence of an event in the future. As a result of the findings, recommendations may include improving:
- Workstation design and process flow
- Chemical handling
- Preventive maintenance
- Equipment
- Policies or procedures
- Employee training

RECORDS

All accident reports should be maintained on file permanently and are discoverable. They should receive timely review by upper management to ensure that proper corrective actions have been taken. Consider the following:
- Just the facts . . . no opinions!
  - Analyze data.
  - Determine the causes.
  - Develop corrective action recommendations.
  - Determine why the accident happened.
- Written report including
  - Date, time, location
  - People—those injured and witnesses who were involved
  - Activities just prior to the accident
  - Accident's sequence of events
  - Accident results
  - Diagrams and photos
  - Immediate temporary corrective actions taken at the scene
  - Recommended permanent corrective actions

Ethics is a rational reflection upon good and evil. The word *ethics* refers to one's identification of the "good" in any given situation as well as the rationale for that identification.

- It is important to understand ethics and know the professional codes of conduct before being faced with an ethical dilemma.
- Published codes of professional conduct can serve as guidance.
- Being objective helps to keep a steady moral compass. "Objectivity" in professional ethics means understanding that there are principles and values outside of oneself that the members of a certain community share and can discuss and against which the individuals will be measured.
- It helps to seek the advice of colleagues. Ask what a reasonable peer would do in the same situation. Solving ethical problems is clearer when one is informed, well-intentioned, objective, rational, and detached but also empathetic, unbiased, sane, sober, etc.
- Every SH&E professional will need to make decisions that challenge personal ethics.

There are rigorous professional guidelines and regulations regarding ethics for an SH&E professional. Below is a list of some of them:

- American Society of Safety Engineers Code of Professional Conduct
- Board of Certified Safety Professionals Code of Ethics and Professional Conduct
- American Industrial Hygiene Association Code of Ethics
- American Board of Industrial Hygiene Code of Ethics
- International Code of Ethics for Occupational Health Professionals

- Contractor Code of Business Ethics and Conduct (48 *CFR* 3.10)
- American Society of Civil Engineers Code of Ethics
- National Society of Professional Engineers Code of Ethics
- Institute of Hazardous Materials Management Code of Ethics

FUNDAMENTAL SH&E PROFESSIONAL ETHICAL CANONS

1. Hold paramount the safety, health, and welfare of the public.
2. Perform services only in areas of competence.
3. Issue public statements only in an objective and truthful manner.
4. Act for each employer or client as a faithful agent or trustee.
5. Avoid deceptive acts.
6. Conduct oneself honorably, responsibly, ethically, and lawfully so as to enhance the honor, reputation, and usefulness of the profession.
7. If the SH&E professional's judgment is overruled because of circumstances that endanger life or property, he or she shall notify his or her employer or client and such other authority as may be appropriate.
8. Deliver competent services with objective and independent professional judgment in decision making.
9. SH&E professionals shall not reveal facts, data, or information without the prior consent of the client or employer, except as authorized or required by law.
10. An SH&E professional shall not permit the use of his or her name or that of an associate in business ventures with any person or firm that he or she believe is engaged in a fraudulent or dishonest enterprise.
11. Conduct professional relations using the highest standards of integrity and refrain from compromising professional judgment by engaging in conflicts of interest.

12. SH&E professionals shall undertake assignments only when qualified by education or experience in the specific technical fields involved.

13. SH&E professionals may publicly express technical opinions that are founded upon knowledge of the facts and competence in the subject matter.

14. Ensure that a conflict of interest does not compromise the legitimate interests of a client, employer, employee, or the public and does not influence or interfere with professional judgments.

15. Provide truthful and accurate representations to the public in advertising, public statements, or representations and in the preparation of estimates concerning costs, services, and expected results.

16. SH&E professionals shall disclose all known or potential conflicts of interest that could influence or appear to influence their judgments or the quality of their services.

17. SH&E professionals shall not accept compensation, financial or otherwise, from more than one party for services on the same project, or for services pertaining to the same project, unless the circumstances are fully disclosed and agreed to by all interested parties.

18. Consultants shall not solicit or accept financial or other valuable consideration, directly or indirectly, from outside agents in connection with the work for which the consultants are responsible.

19. Disclose to clients or employers significant circumstances that could be construed as a conflict of interest or an appearance of impropriety.

20. Serve as an agent and trustee, and avoid any appearance of a conflict of interest.

21. Consultants shall not falsify their qualifications or permit misrepresentation of their or their associates' qualifications. They shall not misrepresent or exaggerate their responsibility in or for the subject matter of prior assignments. Brochures or other presentations incident to the solicitation of employment

shall not misrepresent pertinent facts concerning employers, employees, associates, joint venturers, or past accomplishments.

22. Refrain from offering or accepting significant payments, gifts, or other forms of compensation or benefits in order to secure work or that are intended to influence professional judgment.

23. Follow appropriate health and safety procedures in the course of performing professional duties to protect clients, employers, employees, and the public from conditions where injury and damage are reasonably foreseeable.

The primary focus of the Safety, Health and Environmental (SH&E) profession is prevention of harm to people, property, and the environment. SH&E professionals apply principles drawn from disciplines such as engineering, education, psychology, physiology, enforcement, hygiene, health, physics, and management. They use applicable methods and techniques of loss prevention and loss control. "SH&E science" is a term for everything that goes into the prevention of accidents, illnesses, fires, explosions, and other events that harm people, property, and the environment.

In whatever ways people run the risk of personal injury or illness, there are likely to be safety professionals at work studying those risks. SH&E professionals use a wide variety of management, engineering, and scientific skills to prevent human suffering and related losses. Their specific roles and activities vary widely, depending on their educations, experiences, and the types of organizations for which they work. The standard "Criteria for Establishing the Scope and Functions of the Professional Safety Position" (ANSI/ASSE-Z590.2-2003) sets forth common and reasonable parameters of the professional safety position.

FIND MENTORS, BE A LIFELONG LEARNER

Seek out several mentors and give back to the profession by mentoring others. Professional development can facilitate a commitment to lifelong learning. The future SH&E professional must be:

- Ahead of the technology curve, working on the research and development side to reduce risk to near zero at the source.
- Innovation-focused and highly process driven.
- Highly knowledgeable in the fields of leadership,

global medicine, economics, ergonomics, forensics, industrial engineering and performance excellence, law, sociology, biology, materials science, robotics, computer science, finance, communications, physics, chemistry, statistics, structural and civil engineering, systems safety, environmental/safety management systems, business development, cultural ethics, and, of course, psychology.

- Synchronized with world events and how they may impact SH&E implementation performance.
- Superflexible and highly receptive and adaptive to change.
- Able to demonstrate leadership and drive positive SH&E culture.

CERTIFICATES AND CERTIFICATIONS

Approximately 300 SH&E certification programs and titles are available in the United States. The large number of titles creates many questions and sometimes confusion for potential candidates of these programs: What characteristics define quality? How do safety and environmental certifications add value? What makes one safety certification more worthwhile than another? These are all important questions one must consider when pursuing professional credentialing. The Board of Certified Safety Professionals (BCSP), the American Board of Industrial Hygiene (ABIH), and the Institute of Hazardous Materials Managers (IHMM) are examples of the gold standards for certifications.

Reputable certificate programs include:
- Advanced Safety Certificate by the National Safety Council, www.nsc.org
- Certificate in Safety Management by the ASSE, www. asse.org

Certification	Certificate
Results from an assessment process	Results from an educational process
Typically requires some amount of professional experience	For newcomers and experienced professionals alike
Awarded by a third-party, standard-setting organization	Awarded by training and educational programs or institutions
Indicates mastery/competency as measured against a defensible set of standards, usually by application or exam	Indicates completion of a course or series of courses with a specific focus and is different from a degree-granting program
Standards set through a defensible, industry-wide process (job analysis/role delineation) that results in an outline of required knowledge and skills	Course content set a variety of ways (by faculty committee; dean; instructor; occasionally through defensible analysis of topic care)
Typically results in a designation to place after one's name; may result in a document to hang on a wall or keep in a wallet	Usually listed on a résumé detailing education; may result in a document to hang on a wall
Has ongoing requirements to maintain; individual must demonstrate knowledge of content and that he/she continues to meet requirements	Is the end result; individual may or may not demonstrate knowledge of course content at the end of a set time period

Accreditation confirms that professional credentials meet standards for:
- Validity of the certification examination program
- Fairness of the procedures for determining applicant eligibility
- Adequacy of requirements for ensuring maintenance and enhancement of professional qualifications (recertification)

- Professionalism and independence of the certifying body
- Openness of the program to public scrutiny

These organizations evaluate peer-certification boards for compliance with national and international standards. Many professions recognize the need for certification to establish competency in their respective fields.
- Council of Engineering and Scientific Specialty Boards (CESB)
- American Society for Testing and Materials (ASTM) E1929-98, *Standard Practice for Assessment of Certification Programs for Environmental Professionals: Accreditation Criteria*
- American National Standards Institute (ANSI) ISO 17024
- National Commission for Certifying Agencies (NCCA)
- Accreditation Board of Engineering Technology (ABET)
- Applied Science Accreditation Commission (ASAC)

The Value of Certification
- Public/governmental recognition
- Peer recognition
- Securing employment
- Professional enlightenment
- Professional development
- General
    - Raises bar
    - Levels playing field
    - Provides benchmark
    - Demonstrates competency
- Employers
    - Prescreens candidates
    - Public image
    - Indicator of professionalism
- Safety professionals

- ○ Personal fulfillment
- ○ Peer recognition
- ○ Increased pay and position
  - – CSP makes > 30% more than noncertified professional (2013 survey)
- ○ Competitive advantage
- ○ Demonstrates credibility
- Government agencies
  - ○ Contract qualifications
  - ○ Task performance qualifications
  - ○ Higher public assurance of competency

The top 10 SH&E professional certifications:

1. CSP—Certified Safety Professional, www.bcsp.org
2. CIH—Certified Industrial Hygienist, www.abih.org
3. CHMM—Certified Hazardous Materials Manager, www.ihmm.org
4. CFPP—Certified Fire Protection Professional, www.nfpa.org
5. OHST—Occupational Health Safety Technologist, www.bcsp.org
6. CHST—Construction Health and Safety Technologist, www.bcsp.org
7. CET—Certified Environmental, Safety and Health Trainer, www.bcsp.org
8. STS—Safety Trained Supervisor, www.bcsp.org
9. COHN—Certified Occupational Health Nurse, www.abohn.org
10. CMSP—Certified Mine Safety Professional, www.ismsp.com

Certification is an objective, measurable way of determining professional competency that substantiates a professional's specialty knowledge and expertise in his/her field. Certified professionals gain advanced, market-relevant skills that employers recognize and respect and also gain opportunities to connect with a global community of other certified professionals.

Certified SH&E professionals demonstrate their:

- **Professionalism:** Certification indicates a high level of professionalism to co-workers and customers, thus increasing one's value in the workplace.
- **Leadership:** Certification signifies a dedication to continuous improvement.
- **Recognition:** Certification is an indication of competency, which can enhance one's career path.
- **Knowledge:** Certification improves a professional's understanding of the most current processes and trends in the field.
- **Ethical behavior:** Certification requires adherence to a code of ethics appropriate to the profession.

There are key features to consider when choosing a professional association. These include:

1. **Information.** "I want to receive accurate and prompt information about the issues I need to know about." Better information means professionals can make better decisions that affect business practices.

2. **Public relations.** "I want an association that actively markets to the general public and government on a consistent basis regarding the scope and benefits of the SH&E profession's services and that counters any negative press that may harm the profession." Better educating employers, government, industry, and professionals themselves about the value of the SH&E profession and professionals helps to secure a solid foundation and counter questions related to its value.

3. **Professional development.** "I want high-caliber, international experts brought to local association events so that I can learn directly from the masters." Better resources to enhance lifelong learning are critical to the survival of the profession and to preparing the professional for the ever-changing landscape within the SH&E industry.

4. **Opportunities for involvement.** "I want to be engaged beyond paying dues and feel like I can participate on any level I have time for and choose." There are many opportunities, from committees to working groups, that have defined tasks to be completed over time to create more value for the profession and the professional membership organization.

Being active in an association for an extended length of time has many advantages, including the following:

- Talking to others in your profession helps keep you current on industry trends, products, services, and technologies.

- Belonging to an association provides opportunities to find incredible mentors and to give back by mentoring others.
- Listening to a variety of speakers with diverse backgrounds greatly enhances your overall knowledge of your profession.
- Brainstorming with the same people, month in and month out, allows you to develop solid relationships within your profession by offering both consensus and contradicting views that can lead to better solutions.
- Learning about potential job opportunities/career advancement that otherwise would have been outside your network helps you progress in your career.
- Getting involved in professional meetings allows you to practice leadership among your peers by heading up an event or chairing a committee.
- Actively serving on the board of an association can enhance your professional brand and status within your industry.
- Being on the board also allows you to give back to the organization by volunteering your time and helping others. This level of involvement will enrich your professional development and, to a certain degree, contribute to the overall betterment of the world through your involvement in an organization dedicated to preserving our collective natural resources.

Having developed long-term professional contacts through effective networking, you can call on your contacts for advice if you have a technical question or are trying to learn a new technology.

Many organizations have been formed to further the knowledge base in their industry or specialty. Increasingly, these organizations have developed Web sites to publicly share information. Several of these safety-related organizations and associations are listed here, but many more exist.

Alliance of Hazardous Materials Professionals, http://ahmpnet.org/

American Association of Occupational Health Nurses, www.aaohn.org

American Board of Industrial Hygiene, www.abih.org

American Conference of Governmental Industrial Hygienists, www.acgih.org

American Industrial Hygiene Association, www.aiha.org

American National Standards Institute, www.ansi.org

American Society for Testing and Materials, www.astm.org

American Society of Safety Engineers, www.asse.org

Board of Certified Safety Professionals, www.bcsp.org

Bureau of Labor Statistics, www.bls.gov

Canada Safety Council, www.safety-council.org

Canadian Center for Occupational Health and Safety, www.ccohs.ca

Canadian Registration Board of Occupational Hygienists, www.crboh.ca

Canadian Society of Safety Engineering, www.csse.ca

Centers for Disease Control, www.cdc.gov

Construction Safety Council, www.buildsafe.org

Consumer Product Safety Commission, www.cpsc.gov

Department of Transportation, www.dot.gov

Environmental Protection Agency, www.epa.gov

Food and Drug Administration, www.fda.gov

Health Physics Society, www.hps.org

Human Factors and Ergonomics Society, www.hfes.org

Insurance Institute of America, www.aicpcu.org

International Standardization Organization, www.iso.org

National Fire Protection Association, www.nfpa.org

National Institute of Occupational Safety and Health, www.cdc.gov/niosh

National Institutes of Health, www.nih.gov

National Safety Council, www.nsc.org

Occupational Safety and Health Administration, www.osha.gov

Performance Based Safety, www.safetyconsultants.org

Underwriters Laboratories (UL), www.ul.com

American Board of Industrial Hygiene (ABIH). *American Board of Industrial Hygiene Code of Ethics*, May 2007. Retrieved July 11, 2013.

ASSE/ANSI Z490.1-2012 standard on *Criteria for Accepted Practices in Safety, Health and Environmental Training*

American Conference of Governmental Industrial Hygienists (ACGIH). *2013 TLVs® and BEIs®.* Cincinnati, OH: Author, 2013.

American National Standards Institute (ANSI). *American National Standards Catalog.* Washington, DC: Author, 2014.

American Society of Civil Engineers (ASCE). *ASCE Code of Ethics*, October 2009. Retrieved July 11, 2013.

Board of Certified Safety Professionals (BCSP). *Career Paths in Safety.* Savoy, IL: Author, 2007.

——. *Certified Safety Professional Candidate Handbook.* Savoy, IL: Author, 2014.

——. *Code of Ethics and Professional Conduct.* Savoy, IL: Author, January 2013. Retrieved July 11, 2013.

Board of Environmental, Health & Safety Auditor Certifications. *Performance and Program Standards for the Professional Practice of Environmental, Health & Safety Auditing.* Altamonte Springs, FL: Author, 2008.

Brauer, R. *Safety and Health for Engineers.* 2nd ed. New Jersey: Wiley & Sons, 2006.

Finucane, E. *Definitions, Conversions, and Calculations for Occupational Safety and Health Professionals.* 3rd ed. Boca Raton, FL: Lewis Publishers, 2006.

Manuele, F. *On the Practice of Safety.* 4th ed. New Jersey: Wiley and Sons, 2013.

National Fire Protection Association (NFPA) Fire Protection Handbook 20th Edition 2008 Quincy, MA: Author, 2008.National Fire Protection Association (NFPA). *NFPA 10—Portable Fire Extinguishers.* Quincy, MA: Author, 2007.

——. *NFPA 13—Installation of Sprinkler Systems.* Quincy, MA: Author, 2013.

——. *NFPA 30—Flammable and Combustible Liquids Code.* Quincy, MA: Author, 2012.

——. *NFPA 58—LP Gas Code.* Quincy, MA: Author, 2011.
NFPA 68—Guide for Venting of Deflagrations Quincy, MA Author, 2013

NFPA 69—Standard on Explosion Prevention Systems Quincy, MA Author, 2014

——. *NFPA 70—National Electrical Code.* Quincy, MA: Author, 2011.

NFPA 80—Standard for Fire Doors and Other Opening Protectives, Quincy, MA Author, 2013

——. *NFPA 72—National Fire Alarm Code.* Quincy, MA: Author, 2007.

NFPA 77—Recommended Practice on Static Electricity Quincy, MA Author, 2014

NFPA 80—Standard for Fire Doors and Fire Windows
Quincy, MA Author, 2013

———. *NFPA 101—Life Safety Code.* Quincy, MA: Author,
2012.

NFPA 220—Standard of Types of Building
Construction Quincy, MA Author, 2012

NFPA 251—Standard Methods of Fire Tests of
Fire Endurance of Building Construction and
Materials Quincy, MA Author, 2006

NFPA 495—Explosive Materials Code Quincy, MA
Author, 2013

———. *NFPA 704—Standard for the Identification of
Fire Hazards of Materials for Emergency Response.*
Quincy, MA: Author, 2010.

National Institute of Occupational Safety and Health
(NIOSH). *Pocket Guide to Chemical Hazards.*
Washington, DC: U.S. Government Printing Office,
2013.

National Safety Council. *Accident Prevention Manual for
Business & Industry: Administration & Programs,
Engineering & Technology.* 13th ed. Itasca, IL: Author,
2009.

Petersen, D. *Techniques of Safety Management: A Systems
Approach.* 4th ed. Des Plaines, IL: ASSE, 2003.

Plog, B. *Fundamentals of Industrial Hygiene.* 5th ed.
Itasca, IL: National Safety Council, 2002.

Snyder, D. *Certified Environmental Safety & Health
Trainer Exam Study Guide.* Nixa, MO: Performance
Based Safety LLC, 2014.

Snyder, D. *Certified Hazardous Materials Manager Self Study Workbook*. Nixa, MO: SPAN International Training LLC, 2014.

Snyder, D. *Certified Safety Professional Self Study Workbook*. Nixa, MO: SPAN International Training LLC, 2014.

Snyder, D. *The Hazardous Materials Management Desk Reference*. 3rd ed. Bethesda, MD: Alliance of Hazardous Materials Professionals, 2013.

Snyder, D. *Safety Trained Supervisor Exam Study Guide*. Nixa, MO: Performance Based Safety LLC, 2014.

U.S. Department of Labor. *Code of Federal Regulations*. 29 *CFR* 1910. Washington, DC: U.S. Government Printing Office, 2014.

U.S. Department of Transportation *Code of Federal Regulations*. 49 *CFR*. Washington, DC: U.S. Government Printing Office, 2014.

——. *Code of Federal Regulations*. 29 *CFR* 1926. Washington, DC: U.S. Government Printing Office, 2014.

US DOT Emergency Response Guidebook, 2012 Edition

Yates, D. *Safety Professionals Reference and Study Guide*. Boca Raton, FL: CRC Press, 2011.